ASSERTION-BASED DESIGN

ASSERTION-BASED DESIGN

Harry Foster
Verplex Systems, Inc.

Adam Krolnik
LSI Logic Corporation.

David Lacey
Hewlett-Packard Company

Kluwer Academic Publishers
Boston/Dordrecht/London

Distributors for North, Central and South America:
Kluwer Academic Publishers
101 Philip Drive
Assinippi Park
Norwell, Massachusetts 02061 USA
Telephone (781) 871-6600
Fax (781) 871-6528
E-Mail <kluwer@wkap.com>

Distributors for all other countries:
Kluwer Academic Publishers Group
Post Office Box 322
3300 AH Dordrecht, THE NETHERLANDS
Telephone 31 78 6576 000
Fax 31 78 6576 474
E-Mail <orderdept@wkap.nl>

Electronic Services <http://www.wkap.nl>

Library of Congress Cataloging-in-Publication

CIP info or:

Title: Assertion – Based Design
Author: Harry Foster, Adam Krolnik and David Lacey
ISBN: 1-4020-7498-0

Printed on acid-free paper.

Printed in the United States of America

Dedicated to:

Jeanne—the most wonderful person in my life. And to my children—Elliott, Lance, and Hannah. Always remember, when I was your age I used to have to walk across the room to change the channel.

-Harry

Cindy, Seth, Nicholas, Sarah and Jesus the Christ.

-Adam

To my loving wife, Deborah, for her patience and support while this book was written.

-David

TABLE OF CONTENTS

Chapter 7 Assertion Cookbook 225

Appendix A Open Verification Library 285

FOREWORD

There is much excitement in the design and verification community about assertion-based design. The question is, who should study assertion-based design? The emphatic answer is, both design and verification engineers.

What may be unintuitive to many design engineers is that adding assertions to RTL code will actually reduce design time, while better documenting design intent.

Every design engineer should read this book! Design engineers that add assertions to their design will not only reduce the time needed to complete a design, they will also reduce the number of interruptions from verification engineers to answer questions about design intent and to address verification suite mistakes. With design assertions in place, the majority of the interruptions from verification engineers will be related to actual design problems and the error feedback provided will be more useful to help identify design flaws. A design engineer who does not add assertions to the RTL code will spend more time with verification engineers explaining the design functionality and intended interface requirements, knowledge that is needed by the verification engineer to complete the job of testing the design.

Every verification engineer should read this book! The smart verification engineer will assist the design engineer to add assertions to the RTL-design code because the sooner a design engineer understands the usage and benefits of inserting assertions into the design, the more valuable that design engineer will be to the verification effort. A smart verification engineer is someone who can help a designer to catch the vision and understand the ease and value of assertion-based design. This is the first book to comprehensively address and explain HDL assertion-based design.

My colleague Harry Foster is the best-known name in the Verilog verification and assertion-based methodology community. Along with Lionel Bening, Harry pioneered the Verilog Open Verification Library (OVL), a freely available set of verification-focused Verilog modules that have been used in advanced design and verification environments ever since they were introduced.

My colleague Adam Krolnik was the verification champion of the Verilog-2001 Standards Group. I counted on Adam to promote and propose verification enhancements to the IEEE Verilog language.

David Lacey, Harry and Adam are key participants on the Accellera SystemVerilog Standards Group. Their practical verification experience has contributed to the value of the assertion enhancements added to the SystemVerilog standard.

These three verification specialists have written a book that will endow the reader with an understanding of the fundamental and important topics needed to comprehend and implement assertion-based design.

Included in Chapter 7 of this book is a valuable set of commonly used assertion examples to help the reader become familiar with the capabilities of assertion-based design. This book is a must for all design and verification engineers.

Clifford E. Cummings

Verilog Guru & President, Sunburst Design, Inc.
Member IEEE 1364-1995 Verilog Standards Group
Member IEEE 1364-2001 Verilog Standards Group
Member IEEE 1364-2002 Verilog RTL Synthesis Standards Group
Member Accellera SystemVerilog 3.0 Standards Group
Member Accellera SystemVerilog 3.1 Standards Group

PREFACE

You may have heard that this is a book about verification and now you're wondering why it's called *Assertion-Based Design*, and not *Assertion-Based Verification*. The answer to that is one of the driving forces in this book: Verification doesn't happen in a vacuum. Specification has to occur before any form of verification, and as you know, specification occurs very early in the design cycle. Thus, our contention is that assertion specification is one of the integral pieces of a contemporary design cycle.

Within that context then, the focus of this book is three-fold:

- How to specify assertions
- How to create and adopt a methodology that supports assertion-based design (predominately for RTL design)
- What to do with the assertions and methodology once you have them

To support these three over-arching goals, we showcase multiple forms of assertion specification: Accellera Open Verification Library (OVL), Accellera Property Specification Langauge (PSL), and Accellera SystemVerilog.

The recommendations and claims we make in this book are based on our combincd actual cxpcricnccs in applying an assertion-based methodology to real design and verification as well as our work in developing industry assertion standards.

Real-world experience. In *Assertion-Based Design*, we have pooled our combined experiences to share our understanding and provide a reality-based picture of our chosen topic. The following is a summary of our background related to this topic:

- Harry Foster—Chairs the Accellera Formal Verification Technical Committee; which is developing the PSL standard; created the Open Verification Library; member of the SystemVerilog Assertion Committee; and previously developed assertion-based methodologies at Hewlett-Packard Company.

- Adam Krolnik—Accellera SystemVerilog Assertion Committee committee member and major contributor in the development of the SystemVerilog assertion constructs; created assertion-based methodologies at Cyrix and LSI Logic Corporation.
- David Lacey—Chairs the Accellera Open Verification Library committee; member of the SystemVerilog Assertion Committee; created functional coverage and assertion-based methodologies at Hewlett-Packard Company.

Fundamentals. Property specification is fundamental to an assertion-based verification platform (that is, assertions, constraints, and functional coverage). Once specified, properties enable the following components, which may be included in your assertion-based verification platform:

- *verifiable testplans* through property specification (for example, executable *functional coverage models*, which help answer the question "*what functionality has not been exercised?*")
- *exhaustive* and *semi-exhaustive* formal property checking technology (for example, model checking and bounded-model checking)
- *dynamic property checking* technology (for example, monitoring assertions in simulation) for improved observability to reduce the time involved in debug
- *hardware verification languages* (HVLs) for testbench generation that leverage property specification to define expected input (constraints) and output (assertions) behavior
- *constraint-driven stimulus generation* based on interface properties targeting block-level designs
- *assertion property synthesis* to address silicon observability challenges during chip bring-up in the lab, as well as operational error detection required for high availability (HA) class systems

In this book, we discuss the important role that property specification plays in an assertion-based verification flow.

Evolution in levels of abstraction. The following figure shows an evolution in levels of design notation and specification abstraction. Each time we move up a level of abstraction, we expand possibilities, increase productivity, and improve communication of design intent. Perhaps most importantly, is the growth our field has experienced in conceptualizing new forms of specification, developing new technologies based on these new forms of specification, and then developing standards, which in turn opens new markets. For example, the development of Register Transfer Languages in the mid-1960's lead to the development of synthesis. However, it was the standardization of

VHDL/Verilog in the early 1990's that opened new markets and helped drive synthesis adoption.

The way design and verification has traditionally been performed is changing. In the future, we predict that design and verification will become property-based. Through the standardization of assertion and property languages that are occurring at the time of this publication, we foresee new and exciting EDA markets emerge, once again opening the door for improved productivity. All made possible through assertion-based design practices.

Design notation and specification levels of abstraction

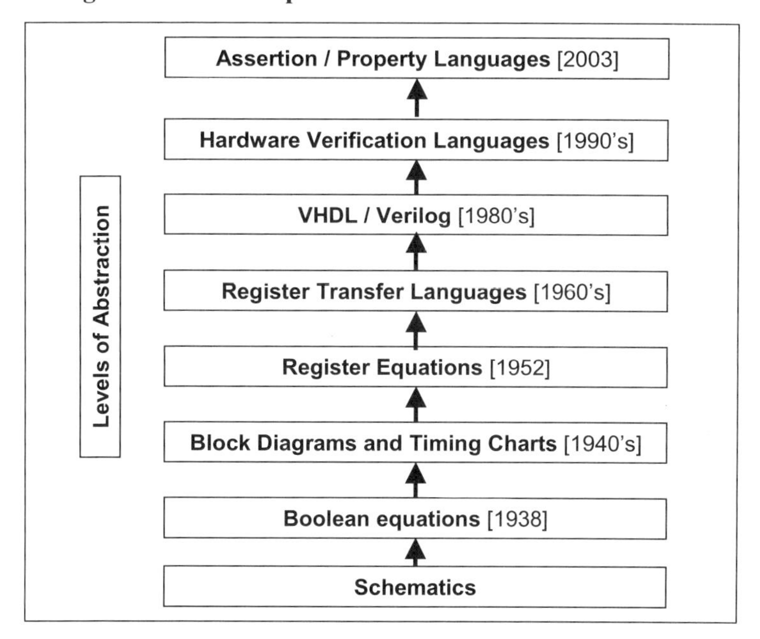

Book organization

Assertion-Based Design is organized into chapters that can be read sequentially for a full understanding of our topic. Or you may wish to focus on a particular area of interest and visit the remaining chapters to supplement your understanding. We have allowed repetitions to support the readers who prefer to browse for information as well as readers who intend to read cover-to-cover and then revisit sections of particular personal relevance. Highlights of the contents follow.

Chapter 1, "Introduction"

Chapter 1 introduces property checking and modern verification techniques. It then introduces assertions and their use in industry, and statistics of their effectiveness. It discusses the benefits of assertion-based verification methodologies and dispels common misconceptions about assertion use within an RTL design flow.

Chapter 2, "Assertion Methodology"

Chapter 2 discusses details for effectively creating and managing an assertion-based methodology, which includes validating assertions using simulation and formal verification. It shows how to best apply assertion-based methodologies to both new and existing designs. It explores how an assertion-based methodology works together with, not separate from, a general design methodology. Finally, it introduces several key learnings that must be accepted by the project team in order to be successful with an assertion-based methodology.

Chapter 3, "Specifying RTL Properties"

Chapter 3 discusses basic property and assertion specification concepts and definitions. It introduces various forms of RTL assertion specifications, which include: the Accellera Open Verification Library (OVL), Accellera PSL formal property language, and Accellera SystemVerilog assertion constructs. Examples are given throughout the chapter to demonstrate how each property concept is used in concert with one of the assertion specification forms to create assertions. Many of these assertion examples can be used today, as they utilize the OVL. Other examples can be used as the EDA vendors move to support PSL and SystemVerilog. Finally, the chapter closes with a PCI property specification example.

Chapter 4, "PLI-Based Assertions"

Chapter 4 discusses PLI-based assertion and procedural assertion techniques. It provides actual code that can be used as an assertion specification form with today's Verilog simulators that support the PLI. It also addresses ways to prevent false firing of procedural assertions.

Chapter 5, "Functional Coverage"

Chapter 5 introduces the verification coverage models and how this topic fits in a book about assertions. It discusses the concepts of black-box and white-box testing, as well as assertion techniques that improve functional coverage. It explores the ideas of controllability and observability and their role in verification. The chapter focuses primarily on functional coverage models, but introduces other traditional coverage metrics (such as programming code coverage) and their role in the overall coverage efforts. Real world examples of successes using functional coverage is provides in this chapter as well as a discussion of the benefits of using functional coverage in your verification efforts. It provides an outline for an effective functional coverage methodology, including guidelines for achieving correct functional coverage density within your design. Finally, it provides a description of several forms of functional coverage specification before concluding with examples of both low-level RTL implementation and high-level architectural functional coverage specification.

Chapter 6, "Assertion Patterns"

Chapter 6 introduces the concept of assertion patterns as a convenient method to document and communicate commonly occurring assertions that are found in today's RTL designs. This provides an easy reference for broad classes of assertions that allows the reader to more easily apply the concepts to their specific designs. A number of different assertion patterns are presented, including signal, set, conditional, past and future event, window, and sequence. Each assertion pattern is described in great detail and provides examples of applicable assertions and descriptive waveforms for additional clarity, as needed.

Chapter 7, "Assertion Cookbook"

Chapter 7 provides concrete examples of the common assertions and functional coverage points for components, interfaces, and general logic, including queues, stacks, finite state machines, encoders, decoders, multiplexers, state table structures, memory, and arbiters. It includes examples that use each form described in Chapter 2 (OVL, SystemVerilog, and PSL). While the list of components is not exhaustive, the chapter gives a broad coverage of components that enables readers to extend the concepts to their specific applications.

Appendix A, "Open Verification Library"

This appendix provides a detailed discussion of the most commonly used monitors in the Open Verification Library.

Appendix B, "PSL Property Specification Language"

This appendix provides a detailed discussion of the most commonly used keywords and operators in the PSL property specification language.

Appendix C, "SystemVerilog Assertions"

This appendix provides a detailed discussion of the most commonly used keywords and operators in the proposed SystemVerilog standard.

Acknowledgements

The authors wish to thank the following people who participated in discussions, made suggestions and other contributions to our *Assertion-Based Design* project:

Salim Ahmed, Tom Anderson, Roy Armoni, Lionel Bening, Dino Caporossi, Michael Chang, KC Chen, Carina Chiang, Edmond Clarke, Claudionor Coelho, Cliff Cummings, Bernard Deadman, Surrendra Dudani, Cindy Eisner, Jeffrey Elbert, E. Allen Emerson, Tom Fitzpatrick, Peter Flake, Gary Gostin, Ramin Hojati, Faisal Haque, John Havlick, Bert Hill, Richard Ho, Tony Jones, Yaron Kashai, Avner Landver, James Lee, Amir Lehavot, Andy Lin, Joseph Lu, Adriana Maggiore, Erich Marschner, Anthony McIsaac, Steve Meier, Hillel Miller, Prakash Narain, Carl Pixley, Andy Piziali, David Price, Jeff Quigley, Vigyan Singhal, Sean Smith, Michal Siwinski, Sandeep Shukla, Bassam Tabbara, Sean Torsney, Andy Tsay, Mike Turpin, David Van Campenhout, Moshe Vardi, Paul Vogel, Tony Wilcox, Yaron Wolfsthal.

Special thanks to Cliff Cummings, Lionel Bening, Avner Landver, Erich Marschner, Andy Piziali, Paul Vogel, Tony Wilcox.

Finally, a very special thanks to Jeanne Foster for providing high quality editing advice and services throughout this project.

CHAPTER 1

INTRODUCTION

Ensuring functional correctness on RTL designs continues to pose one of the greatest challenges for today's ASIC and SoC design teams. Rooted in that challenge is the goal to shorten the verification cycle. This requires new design and verification techniques.

In this book, we address the functional correctness challenge within a contemporary verification flow that relies on an *assertion-based methodology* and *property checking* techniques. The methodology we propose enables designers to meet today's aggressive time-to-market goals, while providing higher confidence in functional correctness. It benefits dynamic verification (that is, simulation), while providing a seamless path to static (formal) verification.

This chapter provides a general introduction to property checking and assertion techniques. We present the benefits associated with assertion-based design and address the many fallacies associated with their use. Finally, we discuss the importance of a specification-driven methodology related to design and implementation.

1.1 Property checking

So, what is *property checking*? In general, you can think of a design property as a proposition of expected design behavior (that is, *design intent*). The proposition *only a single tri-state driver on the memory bus is enabled at a time* is an example of a property for a specific tri-state bus within a design. We can then *assert* that the property must hold (that is, evaluate *true*) for our design, and check our assertion using a dynamic (simulation) or static (formal) verification tool. Other examples of design implementation properties include:

- bus contention
- bus floating
- set/reset conflicts
- RTL (Verilog) "don't care" checks
- full case or parallel case assumptions
- clock-domain crossing checks

Emerging static verification tools automatically extract many design properties through structural analysis of the RTL model. These tools attempt to exhaustively verify these properties using formal techniques (see Section 2.5 Assertions and formal verification on page 44 for a discussion on the *state explosion* problem). When successful, this enables the engineer to verify (or debug) many design properties early in the design cycle—without the need to create testbenches and test vectors. However, designs also include properties that are not as obvious as these examples and that cannot be automatically extracted.

Some properties reflect standard design structures used in standard manners. For example, a queue structure normally operates within its bounds; that is, it neither overfills (overflow) nor removes invalid information (underflow). When a design engineer uses this framework to claim that one of the implemented queue structures can never overflow nor underflow, this is a user-defined property that requires validation. However, until recently, the industry lacked a standard way of specifying RTL user-defined properties that multiple verification tools could recognize and use. In this book, we demonstrate assertion techniques that specify RTL user-defined properties using a subset of the following Accellera standards:

- Open Verification Library (OVL)
- PSL Property Specification Language
- SystemVerilog proposed assertion constructs

1.2 Verification techniques

Traditionally, engineers verify the design's implementation against its requirements using a *black-box* testing approach. In other words, the engineer creates a model of the design written in a hardware description language (for example, Verilog [IEEE 1364-2001] or VHDL [IEEE 1076-1993]). The engineer then creates a testbench, which includes or instantiates a copy of the model or *device under verification* (DUV). Historically, *testbenches* would read a vector file as input—and apply the vectors to the DUV cycle-by-cycle. The DUV output results were then compared against a reference model. The ability to directly "observe and validate" came later with the development of *self-checking testbenches.* Recently, testbenches have become complex verification environments often built with a hardware verification language (HVL) that combines:

- automatic vector generation,
- output response validation, and
- coverage analysis.

The specification defines the legal values or sequences of values permitted by the DUV's input and output ports (that is, a black-box view of the design).

One problem encountered when using a black-box testing approach is that the DUV might exhibit improper internal behavior, such as a state machine violating its one-hot property, but still have a proper output response (at a specific, observed point in time). In cases such as this, a design error exists, but it will be missed because it is not directly observable on the output ports. This might be due to the current set of input stimulus, which, when applied to the DUV, impedes the internal problem's value from propagating to an output port. Given a different set of input stimulus (or if the simulation were to run a few clocks longer), the internal error might be observable. However, validating all internal properties of a design using black-box testing techniques is impractical, particularly as design size increases.

Alternatively, *white-box* testing can be implemented to validate properties of a design. This technique adds assertions that monitor internal points within the DUV, and results in an increase in observable behavior during testing. For example, using the DUV described above, we can add an assertion (or monitor) to the design to directly observe and validate whether a state machine is always one-hot. Thus, if the one-hot property is violated, the error is instantly isolated to the faulty internal point. This overcomes the problem associated with black-box testing, which is the possibility of missing an internal error (for a given input stimulus) by observing only the DUV output responses.

1.3 What is an assertion?

In general, an *assertion* is a statement about a design's *intended behavior* (that is, a property), which must be verified. Unlike design code, an assertion statement does not contribute in any form to the element being designed.[1] Its sole purpose is to ensure consistency between the designer's *intention,* and what is *created.*

Consider the following analogy:

A designer issues the print command for a report, walks to the printer expecting to find 11 pages, but finds only 8. *What happened to the missing pages?*

The printer displays the following message:

> ERROR - PAPER JAM AT SECTIONS 3, 7.

In this case, the printer assertion triggered when it detected a difference between the expectation of the user and the creation from the printer. It notified the user of the error and where the potential problems occurred. This analogy demonstrates three key features of assertions:

- *error detection*
- *error isolation*
- *error notification*

Each of these features is discussed in detail throughout the remainder of this book.

1.3.1 A historical perspective

Design verification is a process used to ensure that a circuit or system model (that is, the implementation) behaves according to a given set of requirements (that is, the specification). Typically, the requirements consist of a natural language list of assertions, each of which must be verified. In fact, over 50 years ago, Alan Turing [1950] made the following observation concerning partitioning a large verification problem into set of assertions:

> *How can one check a large routine in the sense of making sure that it's right? In order that the man who checks may not have too difficult a task, the programmer should make a number of definite assertions which can be checked individually, and from which the correctness of the whole program easily flows.*

Over 30 years ago, Floyd [1967] and Hoare [1969] proposed the concept of using formalisms (that is, property specification—or assertions) to specify and reason about software programs. Moreover, software engineers have long used assertions within

1. Note that assertions can be used to constrain the synthesis process. For example, the *full_case* and *parallel_case* synthesis directives are really assertions.

their code to check for consistency (for example, checking for null pointers or illegal array index ranges).

design by contract

More recently, new programming languages (such as Eiffel) have emerged that are based on an underlying theory known as *Design by Contract.* Eiffel "views the construction of a software system as the fulfillment of many small and large contracts between clients and suppliers" [Meyer 1992]. Components written in Eiffel specify pre-conditions (a form of assertions) that the user must satisfy to use the component in an acceptable and reliable manner. Then, post-conditions (also assertions) are specified as a definition of what the component will do and the properties the results will satisfy. The post-conditions assume that the component pre-conditions are satisfied. Systems can become very robust when the operating conditions (pre-conditions) are tightly controlled.

RTL assertions

RTL assertions written for hardware interfaces can achieve the same effect that Eiffel components obtain from their assertions: a controlled environment for interface usage and an understood set of expectations. Consider an intellectual component purchased as intellectual component (IP) from another company. It consists of the component in a usable form, and an instruction manual on how to use the component. If the component uses assertions to define its interface, the component will be used more successfully than if assertions are not present. Additionally, the support effort required by the group or company supplying the IP is reduced because the user is told when they are using the IP incorrectly.

In today's design and verification environment, emerging hardware verification languages include various forms of assertion library templates. Furthermore, hardware description languages (HDLs) include constructs that support assertion specification. For instance, VHDL [IEEE 1076-1993] includes a keyword `assert`, which permits designers to add embedded checkers to model description code. This language construct, as shown in Example 1-1, ensures that any user-specified condition (that is, a Boolean expression) always evaluates to TRUE.

Example 1-1 VHDL assertion syntax

[label] **assert** *VHDL_expression*
  [report message]
   [severity level]

For the VHDL assertion construct, an error is reported when the VDHL_expression evaluates to FALSE. The assertion's optional report clause specifies a message string that will be included in error messages generated by the assertion. In the absence of a report clause for a given assertion, the string "Assertion violation" is the default value for the message string. The VHDL assertion's optional severity clause specifies a severity level associated with the assertion. In the absence of a severity clause for a given assertion, the default value of the severity level is ERROR.

Example 1-2 **VHDL example to check for inverted signals**

```
ASSERT ((a = '1') XOR (b = '1'))
  REPORT "error: A & B must be inverted"
    SEVERITY 0;
```

Unlike VHDL, Verilog [IEEE 1364-2001] does not contain an assertion construct. However, checks can be coded in an explicit fashion as show in Example 1-3.

Example 1-3 **Verilog example to check for equality**

```
always (a or b) begin
  if (a ^ b) begin // not equal
    $display("error: A&B must be equal: %m");
    $finish;
  end
end
```

Many present-day commercial tools provide their own proprietary HDL assertion solutions. Most recently (and perhaps most importantly) the Accellera standards organization has engaged in efforts to unify the industry with a standard for an HDL assertion specification. [Fitzpatrick et al. 2002].

1.3.2 Do assertions really work?

Assertions have been used by many prominent companies, including:

- Cisco Systems, Inc.
- Digital Equipment Corporation
- Hewlett-Packard Company
- IBM Corporation
- Intel Corporation
- LSI Logic Corporation
- Motorola, Inc.
- Silicon Graphics, Inc.

Designers from these companies describe their success with methodologies that incorporate assertions as follows:

- **34%** of all bugs were found by assertions on DEC Alpha 21164 project [Kantrowitz and Noack 1996]
- **17%** of all bugs were found by assertions on Cyrix M3(p1) project [Krolnik 1998]
- **25%** of all bugs were found by assertions on DEC Alpha 21264 project - The DEC 21264 Microprocessor [Taylor et al. 1998]
- **25%** of all bugs were found by assertions on Cyrix M3(p2) project [Krolnik 1999]
- **85%** of all bugs were found using OVL assertions on HP [Foster and Coelho 2001]

From these papers, a common theme emerges: When designers use assertions as a part of the verification methodology, they are able to detect a significant percentage of design failures. Thus, assertions not only enhance a verification methodology; they are an integral component. Assertions are typically written to describe design assumptions or a potential corner case involving a lower-level implementation detail. This complements traditional verification methods, which typically focus on higher levels of abstraction (for example, bus transactions) and rarely attempt to verify specific implementation details (for example, a specific state machine is one-hot).

On the Cyrix M3(p2) (3rd gen x86 processor) project sited above, 750 bugs were identified prior to adding assertions. However, the week *after* a significant number of assertions were added to the design, the verification team experienced a three-fold increase in its bug reporting rate. In fact, fifty percent of all remaining bugs were identified through assertions. This represented 25% of all bugs found on the project.

1.3.3 What are the benefits of assertions?

This section explores the benefits of using assertions and their tremendous impact on increasing design quality while reducing the time-to-market and verification costs.

Improving observability

Fundamental to understanding the benefits of using assertions is understanding the concept of *observability*. In a traditional verification environment, a testbench is created to generate stimulus, which is applied to the design model (that is, *design under verification* or DUV). In addition to generating input stimulus, the testbench validates proper output behavior by monitoring (that is, observing) the DUV's output ports. In order to identify a design error using this approach, the following conditions must hold:

1. Proper input stimulus must be generated to activate (that is, sensitize) a bug,
2. Proper input stimulus must be generated to propagate all effects resulting from the bug to an output port.

It is possible, however, to set up a condition where the input stimulus activates a design error that doesn't propagate to an observable output port. In these cases, the first condition cited above applies; however, the second condition is absent.

A benefit of assertions embedded in the code is that they increase the *observability* within the design. In this way, the verification environment no longer depends on the second condition listed above to identify bugs. Thus, any improper or unexpected behavior can be caught closer to the source of the bug.

The experiences of the companies cited above (and others) show that embedded assertions, when added to an existing verification environment, identify problems that previously were not identified. In fact, both the DEC Alpha team [Kantrowitz and Noack 1996] and Cyrix M3 team [Krolnik 1998] demonstrated that when they added assertions after a point when they had assumed simulation was complete, they found additional bugs using the same set of tests that previously passed.

Reducing debug time

isolate bugs

As we previously stated, assertions improve observability, thus enabling us to find bugs exactly when (that is, at a specific time) and where (that is, at a specific location) they occur. Conversely, traditional verification methods do not detect bugs directly in (or close to) the logic that generated them; they typically detect a bug multiple clocks after its occurrence and at some other distant location in the design. The problem with this approach is that it typically requires that multiple designers back-trace multiple paths (or blocks) within the design. This consumes many debug hours before the problematic code is finally identified. When a team is unsure where the bug originated, unpredictable delays ensue as teams pass a failure between several designers for analysis, involve more individuals, and possibly engage in "pointing fingers" and throwing the problem to one another. Isolating bugs closer to the actual source provides a huge time savings and reduces the total resources required to isolate the bug. Thus, it improves projects' time-to-market goals.

The following case illustrates how assertions help isolate bugs by detecting them in the logic that generated them. In this case, the

Cyrix M3 project experienced a test failure that appeared as a time-out to complete an instruction. The flow of failure analysis follows:

- Designer A of the completion logic determined that the memory operation portion of the instruction did not complete.
- Designer B of the data cache controller determined that address translation for the operation was wrong.
- Designer C of the address translation controller indicated the translation was successful and passed it back to designer B.
- Designer B reviewed the simulation, and then realized that he received a wrong translation earlier, and passed the problem back to designer C.
- Designer C reviewed the problem, and then noticed that his code had returned two translations for the preceding translation request from designer B's code.
- Designer B then explained that the extra translation caused the time-out

The actual situation described above occupied most of the day for these three engineers. If designer B had written an assertion to ensure each request was followed by only one completion, the failure would have been identified closer to the problem (the previous instruction) and the actual unit (the translation unit). And, the problem would not have drawn in two of the three engineers. After adding the assertion, the engineers easily fixed the logic and quickly validated the new code.

Improving integration through correct usage checking

check interfaces

Assertions also provide benefits when developing and integrating intellectual property (IP) components. A design team initially validates the IP with a given set of functional constraints and inserts boundary assertions to monitor correct interface communication during integration verification. These boundary assertions form *verifiable contracts* between the IP provider and the IP integrator through *correct usage detection*, a form of self-checking code. When a boundary assertion identifies incorrect usage, it relieves the IP provider of the burden of debugging someone else's design, while enabling the IP user to quickly identify code that doesn't satisfy the IP's specified constraints. In this respect, the IP code is self-checking. That is, it relies on assertions to identify the source of bugs that occur along the input and output boundaries of the IP code.

Improving verification efficiency

find bugs faster

Saving time is perhaps the most significant benefit designers realize with assertions. Experience demonstrates that assertions can save up to 50 percent of debug time, thus reducing overall time-to-market [Abarbanel et al. 2000]. Design teams save debug time when an engineer does not have to backtrack through large simulation trace files and multiple blocks of logic to identify the exact location of the design failure.

work at all times

Unlike conventional debugging processes involving a designer, assertions embedded in the RTL source code are always monitoring for valid (or invalid) behavior. For example, in a conventional debug process, the designer typically goes through the following steps:

- examines a failing testcase,
- identifies the problem,
- fixes the problem, and then
- validates the fix by re-running the testcase that previously failed.

During this process, if designers discover another anomaly associated with the original failing circuit, they will fix it as well. But after the designers find a *fix* for a specific failing testcase (and any other anomalies that emerged in the course of validating the fix), they generally stop looking for other corner cases associated with the bug and move on to the next problem. Unfortunately, there could be a different simulation pattern or sequence, which has not yet been covered, that would identify another corner case associated with the original bug.

Conversely, when designers add assertions to the RTL source, they avoid this situation because assertions never stop monitoring the design for invalid behavior and can help trap many corner case problems during future simulation runs. Hence, a new set of test vectors applied to the design in the future might uncover additional problems associated with a bug they presumed they had fixed.

work with all tools

The assertion-based verification methodology we propose permits the designer to specify assertions in a single form, which then is leveragable across an entire suite of verification tools. In other words, we claim that engineers should only have to specify an assertion once (that is, one way) whether they choose to target the assertions with a custom verification tool, a standard RTL testbench, a commercial simulator, a semi-formal tool, or a formal verification tool.

facilitate formal analysis

Formal specification describing architectural consideration, as well as consistency in protocol design prior to implementation, can be verified using various formal techniques. Once RTL implementation begins, formal technology can be used to explore the design space around assertions within the implementation, further increasing confidence in the final design's ability to function correctly when built. For example, Bentley [2001] reported that 400 bugs were identified and fixed by formally verifying a large set of assertions on a recent Intel Pentium project prior to silicon.

Improving communication through documentation

specify correct behavior unambiguously

In addition to finding bugs, assertions encapsulated in the RTL provide an excellent form of documentation. For instance, a designer may add an assertion to an interface that states the following expected behavior: *a bus request must be followed by a bus grant within five clock cycles*. By adding an assertion for this property, the engineer documents an aspect of the design in a form that is self-checking and easy to convey to other engineers.

Other engineers can review the assertions to understand the low-level specifics of how to interface with another block. Furthermore, assertions formally document protocols, interfaces, and assumptions in an unambiguous form that clarifies a designer's interpretation of the specification and design intent.

1.3.4 Why are assertions not used?

If assertions are such a great enhancement for verification, why aren't engineers using them more extensively in the design process? This section lists some common arguments that are put forth by teams that are not using assertions and identifies the fallacies in reasoning.

- *Where am I going to find time to write assertions? I don't even have time to write comments in my code!*

  Based on our studies and interviews with multiple projects and engineers, the overhead in writing assertions can amount to anywhere between one and three percent extra time added to the *RTL coding phase* of a design project (note that the RTL coding phase is only one of many phases within a project). In general, the overhead is very minor (that is, closer to one percent). Why? Because, regardless whether you write comments or add assertions prior to or during RTL coding, you are already analyzing and thinking about correct or expected behavior of your design. In other words, you are already considering:

- What is the valid interface behavior?
- What are the legal states for my FSM?
- What are the boundary conditions on my queue controller?

By adding assertions, you are formalizing your thought process in a form that is verifiable. In fact, many design errors are avoided prior to RTL coding and verification through this more systematic approach to design and analysis.

- *I have to get my design working first. If there is time later, then I'll add assertions.*

This argument translates to: "Assertions will not save me (us) any time". Yet, experience demonstrates that assertions can save up to 50 percent of debug time [Abarbanel et al. 2000]. These people will change their position when they find that the assertions other engineers used in their blocks effectively isolate failures. When they save debug time just by isolating failures that originate outside their block, they will be convinced of the value of including assertions in their own blocks.

- *I will spend more time debugging the assertions than debugging my code; therefore, they are a waste of time.*

Compare this sentiment to developing testbenches. Does this mean that creating testbenches to verify the design is useless—since the testbenches might also contain bugs? Certainly not, and the same is true with assertions (that is, assertions could contain bugs, but that does not render the practice useless). As with most new experiences, using assertions involves a learning curve. As engineers gain experience with assertions, they recognize what constitutes a correct assertion and spend less time debugging assertions. If engineers are spending time debugging assertions, it is because they did not completely express the assertion, or they did not fully understand some subtle aspect of the design. In fact, the analysis process that takes place while specifying assertions quite often uncovers complex bugs, and this occurs prior to any form of verification.

- *I can't think of any assertions to put in my code. There are no places for them.*

Designers who say they *can't see any potential errors in their code* will be spending a lot of time rewriting their design code. Experienced designers recognize that typographical, transcription, and design errors all contribute to functional problems that must be addressed. Inexperienced designers must be taught how to see the potential for problems. Then, they will see that they can use assertions to naturally check for correctness.

- *The assertions slow down simulations. They waste time.*

During the post-mortem review for a Convex Computer Corporation project[2], a survey was given to a design team with the following question: "If we had computers that could simulate the design 10x faster than today, would that help you debug faster?" One astute designer responded, "No! We run all the tests overnight and I arrive in the morning with a list of failures I can't work through in one day." This illustrates that the debugging process is where efficiency improvements are required. So, although assertions do have an overhead, that overhead is comparable to monitoring the design for correctness. The Cyrix M3 project and the HP ASIC project found that assertions produced an overhead in the range of 15-30 percent [Krolnik 1998] [Foster and Coelho 2001], depending on the number of assertions and amount of monitoring. For example, Cisco reported a 15 percent increase in simulation runtime on a large ASIC project containing 10,656 OVL assertions [Smith 2002].

- *It is difficult to debug with assertions. It's not possible to fail an assertion and continue debugging. [Borland 2002]*

This is actually a critique of the assertion methodology. A good assertion methodology has controls that allow simulations to define an error threshold and controls to determine what action (continue simulation, terminate the simulation, stop, debug) must occur when that threshold is reached. (See Chapter 2, "Assertion Methodology" on page 21 for details on an effective assertion-based verification methodology.)

- *Designers shouldn't check their own code. Hence, adding assertions violates this rule.*

Adding assertions in the RTL design is a way of specifying expected behavior and it is analogous to the following example of asserting that a pointer will never be null in software designs. Software engineers have known for years that it is a good idea to check that a pointer passed into their code is not null prior to use. If a null pointer is encountered during normal execution, the problem can be quickly isolated when an assertion is used. Notice how asserting that a pointer will not be null says nothing about *who* is going to test *what*. In other words, the designer is placing a proposition into the code that states that a particular implementation property will always be TRUE. This is not violating Bergeron's [2000] redundancy verification convergence model (that is, design engineers should not verify their own designs). In fact, the verification engineer will read the design specification and create a set of test scenarios[3] to validate the design. During the course of verification, if a sequence of events emerges in which a calling routine passes a null pointer, then the problem is quickly

2. Based on Foster and Krolnik's experience at Convex Computer Corporation in the early 1990's.

isolated via the implementer's assertion, and this dramatically simplifies debugging. In this respect, RTL design should be no different than software design. When the designer asserts that two signals must always be mutually exclusive, this does not state how the design should be verified. In fact, rarely will the verification engineer focus on implementation-specific testing. Hence, assertions added to the RTL implementation improve the overall quality of the verification process.

1.4 Phases of the design process

The *waterfall refinement* approach to the design process includes three distinct phases: *specify*, *architect/design*, and then *implement*. Although theoretically attractive in principle, this approach is seldom practiced in the real world. In other words, often the engineer moves back and forth between specification, architect/design, and RTL implementation as analysis uncovers additional requirements. In this section, we present an abstract view of these three phases within the overall scope of the design process. Our goal is to demonstrate how assertions help validate lower-level details, which are developed during the architect/ design or RTL implementation phase and are rarely described in the higher-level specification. That is, if the verification process focuses only on validating high-level requirements established in the specification, then there are many details in the design that might go unchecked.

Figure 1-1 illustrate three distinct *regions of design intent* related to a product's development process [Piziali and Wilcox 2002]. This corresponds to the following unique phases of development:

- The *specification phase,* which is the initial step in the design process. In this phase, the architect envisions the design intent and then establishes high-level requirements for the product.
- The *architect/design phase,* which is a process of refining the higher-level intent (described in the specification) into a set of detailed requirements, partitioning the high-level architecture into functional blocks, and considering alternative implementations for each block prior to RTL coding.
- The *RTL implementation phase,* which is the process of coding an RTL implementation such that it satisfies both the high-level specification and design requirements.

3. Or better yet, the verification engineer would implement a functional coverage model of the verification space and design and implement functional and verification constraints in their stimulus generator.

Figure 1-1 **Regions of intent within the design process**

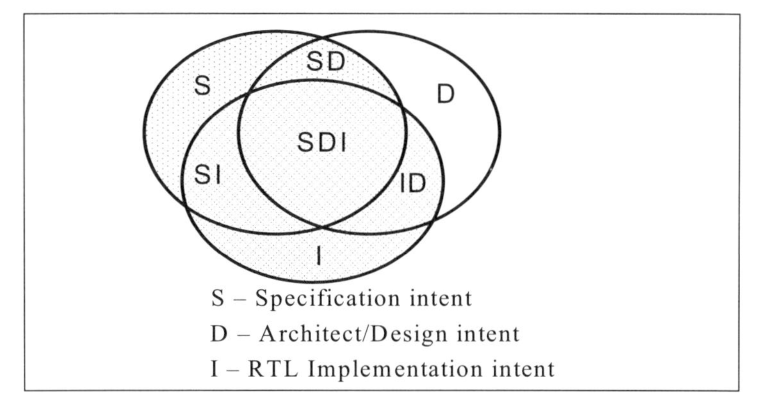

The various regions of intent of the design process, and their overlap in terms of what is actually being specified or designed, are represented as seven domains within the Venn diagram in Figure 1-1:

- S - Specification only. In the worst case, the requirements developed during the specification phase (and contained within this region of the Venn diagram) were for some reason not considered during the architect/design or RTL implementation phases. This generally indicates a serious problem. Hence, ideally this space is very small.
- D - Architect/Design only. This domain consists of design intent details that were not defined in the higher-level specification phase, and for some reason were not implemented in the RTL code. Note that this is either a superfluous part of a design, or missed lower-level functionality during implementation.
- I - RTL implementation only. This domain of intent was not defined by the specification or the architect/design phases. This happens frequently since there are many lower-level implementation details that neither the specification nor the design need to describe.
- SD - Specification and architect/design. This domain was defined by the specification and architect/design phases, but details were either missed during RTL implementation or deemed unnecessary (hence, the requirements were changed but not updated).
- DI - Architect/Design and RTL implementation. The intent details in this domain were defined by the design and RTL implementation phases, but not by the specification phase. This is common, since specifications should not address every detail (allowing enough degrees of freedom during design for optimization).

- SI - Specification and RTL implementation. The intent details in this domain were defined by the specification and RTL implementation phases, but not by architect/design phase. This occurs when the design has errors and is not updated. For example, bug fixes are made that were not covered in the design.
- SDI - Specification, architect/design, and RTL implementation. All requirements, design decisions, and RTL implementation details are covered here. Although the verification engineer's goal is to verify (and ensure consistency and completeness) of all requirements established in the specification phase (represented by the specification domain), it is only the SDI domain that covers all details developed in all three phases of the design process. In other words, any functionality contained within the design or RTL implementation domains that is not contained within the SDI domain would typically not be verified.

In practice, the design process rarely excludes any of these regions of intent. Ideally, some would be minimized while others are maximized. However, the following observations related to Figure 1-1 generally apply to the development process:

1. Designers typically first attempt to merge two of the regions of intent into a single domain (that is, ensure that all higher-level requirements established during the specification phase are satisfied during the architect/design phase) and finally merge the resulting domain with the remaining RTL implementation intent domain (that is, insure that the final RTL implementation satisfies the original requirements of the specification) to achieve higher verification coverage.
2. Each region of intent requires specific techniques for identification and alignment (that is, maximized overlap), which will be discussed in the following sections.

1.4.1 Ensuring requirements are satisfied

the difficult method

Some design methodologies give little attention (or totally ignore) the *architect/design phase*, and instead, immediately begin *RTL implementation* with hopes of meeting all the requirements established by the specification. Thus, in an abstract sense, they attempt to move (that is, merge) the RTL implementation intent with the higher-level specification intent. This is a difficult process to execute. Ignoring the design phase provides little or no benefits typically experienced through a formalized process. This is due to the lack of information sharing between multiple stakeholders in the design process. The designer is essentially working in isolation during the implementation and problem resolution processes. Using this approach, the RTL implementation may suffer long delays (and ultimately poor silicon). Hence, there are better ways to ensure that the RTL implementation satisfies the specification's requirements (that is, in an abstract sense, merging regions of intent).

the more efficient method

A design methodology that utilizes a formalized process provides time to align much of the detailed architect/design requirements with the specification requirements before RTL implementation begins. It also provides opportunities for review and collaborate to ensure the design has the greatest chance of matching the specifications. Once the design has been completed (and reviews are favorable) an implementation can begin. Assertions can play a major role in ensuring that the requirements specified during one phase of design are satisfied during the next phase (that is, aligning the domains). Furthermore, adding design assertions for lower-level details (beyond the higher-level specification) can help capture design intent in a fashion that is both verifiable in the design phase and dramatically simpler to debug during the RTL implementation phase.

3 phases of RTL implementation functional verification

Many people consider the functional verification of an RTL implementation to be a process with a single goal: *identify (and fix) bugs in the implementation.* However, the functional verification process must actually ensure three distinct goals:

1. All lower-level RTL implementation details (outside the requirements established during specification and architect/ design) are bug-free
2. The RTL implementation satisfies all detailed requirements developed during the architect/design phase (which are outside the higher-level specification)
3. The RTL implementation satisfies all properties of its higher-level specification

Meeting these goals is typically performed concurrently during the verification process, although they could be performed in a serial fashion.

An effective way to verify that the RTL implementation matches the design requirements is to write assertions directly in the code during the RTL implementation phase. They immediately identify functional errors in the RTL implementation. As these errors are corrected, the implementation begins to function according to the design intent. In an abstract sense, we are attempting to align the details described by the RTL *implementation* region of intent with the requirements described by the *design/architect* region of intent; that is, ensure that what we have implemented during the RTL coding phase satisfies the detailed requirements we established during the design phase.

Note that for large design errors, it is usually easier to reconsider (that is, return to) the design phase and determine the optimal modification required to address the problem (rather than trying to fix an error directly in the RTL implementation). Ignoring the design phase and considering all problems as implementation errors is tantamount to adjusting the implementation domain to match the specification domain. As previously stated, this is frequently more difficult and (more importantly) requires significantly more time than returning to the design phase for further analysis.

1.4.2 Techniques for ensuring consistency

During the design process, our goal is to ensure that what is implemented in the RTL satisfies the requirements established during the specification phase, as well as the detailed requirements developed during the design phase. This can be viewed as maximizing the overlapping of all regions of intent in Figure 1-1 (or minimizing non-overlapping domains). The following sections list possible techniques for aligning these domains.

specification only

The requirements contained within the specification intent *only* domain (S) point out a serious flaw in our design. The approach to minimize this region is to create a functional coverage model that covers all aspects of the specification. Hence, a good functional coverage model will identify holes in the design, implementation, and verification phases of the design process. That is, a functional coverage model can identify missing functionality during implementation or unverified characteristics of a design.

architect/design only

The architect/design intent only domain (D) might be the result of superfluous architect/design details. However, it may be the result of missing or misunderstood implementation details. Designing functional coverage models (as well as designing reviews against the implementation) is the best way to minimize this domain.

implementation only

The implementation intent only domain (I) occurs frequently due to unspecified RTL implementation details, which were not specified during the specification phase or described during the design phase. Adding assertions to the RTL are necessary during coding to adequately verify the designer's intent. As previously mentioned, this level of detail is rarely known to the verification engineer (and is not contained in the specification), which can result in poor RTL implementation coverage during verification. Code coverage tools are also useful in this region to identify holes in the RTL implementation verification process.

architect/design and implementation

This domain is also common, as details are developed that are not specified during the specification phase. Adding assertions during the design and RTL implementation phase helps verify (and clarify) design intent. Code coverage tools also serve to identify details (outside of the specification) that are not verified within this domain.

specification and architect/ design

This domain represent the case where various higher level requirements (developed during the specification and architect/ design phase) were not implemented. Developing a good functional coverage model helps to identify missing functionality in the RTL implementation (see Chapter 5, "Functional Coverage").

specification and implementation

This domain should be minimal, since it represents details that were not considered during the design phase, yet are necessary for correct functionality. If this domain is large, it may indicate a poor design (that is, little thought was given to design analysis prior to RTL coding).

specification, architect/design and implementation

This is the ideal domain resulting from the three phases of the design process. Assertions derived from the specification, can ensure that the implementation satisfies its requirements. In addition, a good functional coverage model will identify holes in the verification process related to this domain.

1.4.3 Roles and ownership

Specifying intent during the three phases of design generally involves multiple stakeholders, where each stakeholder plays a critical role in the success of the assertion-based methodology. Hence, in this section we outline the various roles and property specification ownership that the architect, verification engineer, and design engineer play in specifying assertions and functional coverage.

- The verification team creates system-level assertions and functional coverage for the *specification intent* (that is, requirements defined in the high-level specification). Assertions, combined with functional coverage, enable the verification team to create a *verifiable testplan.*
- The architect, design leads, or verification team creates block-level assertions, as well as functional coverage, for the *design/ architect intent*. These assertions and functional coverage points are derived from the detailed functional specification, and are used in the testbench or embedded directly in the design model.

- The designer creates assertions and functional coverage for the *RTL implementation intent*. Note that lower-level details of the RTL implementation are rarely the focus of the verification team. Hence, important details could go unverified without the designer's involvement in adding assertions and functional coverage points.

1.5 Summary

An *assertion* is a statement about a design's *intended behavior* (that is, a property), which must be verified. Key verification benefits that assertions provide include:

- *improved error detection*
- *improved error isolation*
- *improved error notification*

These benefits are a result of improved *observability* (when assertions are embedded in the design).

Other benefits include:

- Reduced debug time—up to 50%
- Improved integration through correct usage checking
- Improved verification efficiency through specification
- Improved communication through documentation

In this chapter, we discussed various benefits of adopting an assertion-based design methodology. We also discussed many common arguments that are put forth by teams that are not using assertions and identified the fallacies in their reasoning. Finally, we present an abstract view three phases within the overall scope of the design process (*specify*, *architect/design*, *implement*). We then demonstrated how assertions help validate lower-level details, which are developed during the architect/design or RTL implementation phase and are rarely described in the higher-level specification.

CHAPTER

2

ASSERTION METHODOLOGY

The importance of adopting and adhering to an effective design methodology has been a popular topic of discussion within the engineering community and in literature for many years. It is generally accepted that the benefits of a well-defined design methodology far outweigh the cost of implementing it.

In this chapter, we focus primarily on components of an effective *assertion-based methodology*. However, an assertion-based methodology does not stand alone. In order to reap the greatest benefit, it must be tightly integrated with a larger design methodology. So, this chapter briefly discusses some of the broader design methodologies as a foundation for an effective assertion methodology. We then discuss how to apply an assertion-based methodology targeting *new designs*—followed by a discussion on how to apply an assertion-based methodology targeting *existing designs*. Finally, we discuss simulation-based methodology considerations specifically related to assertions.

Incidentally, an effective assertion-based methodology requires the same up front planning that an effective design methodology requires. This chapter guides the reader through steps that provide a blueprint for this type of comprehensive planning.

2.1 Design methodology

A typical design process includes five basic elements, and these drive our assertion methodology discussion:

1 Project planning
2 Design requirements

3 Design documents
4 Design reviews
5 Design validation

By establishing a consistent approach to each element described in this section, a project team solves many problems that can arise during a project's life cycle. Ignoring any one element greatly contributes to a less effective project (for example, slipping schedule or untested functionality). Your project might even fail by ignoring just one element. The magnitude of the failure may vary, but each element of the design process we recommend is in place to enhance the others and make them successful.

This section briefly discusses each of the basic design process elements and introduces the role of an assertion-based methodology within each of these elements. Our goal is to show that an assertion-based methodology is most effective when it is tightly integrated with the existing design process; that is, it should not be seen as an add-on or an afterthought.

2.1.1 Project planning

The first step to a successful project is planning. A project team must define policies and conventions that describe how to implement, verify, and maintain the project. Additionally, the team must define resources required to successfully complete the project, specify schedules, and outline guidelines that the team will uphold during the design process. These items must work together to make all the tasks easier.

Adequate planning ensures an efficient infrastructure, and documenting the process is an essential step in establishing an effective methodology. This allows the project team to concentrate on getting the job done instead of figuring out *how* to get the job done. An important benefit of establishing policies and conventions is that they allow new members of the team to be brought on-line much quicker, since the plans and resources are available for independent review. In addition, the conventions provide commonality across the design for ease of understanding and later debug of the machine. The conventions guide designers away from constructs with known problems (such as in the areas of timing and correctness). Without defining and documenting policies and conventions in the initial stages of the project, many questions arise and are addressed in an ad-hoc mode, which yields inconsistency across the design team. Finally, establishing project conventions simplifies the task of shuffling people resources between blocks in later stages of design.

Planning covers a wide variety of areas, including those presented below. Each of these should be analyzed and defined.

Project documents

This section sets forth guidelines for project policies, conventions, and documentation. This information may include the following resources:

- Requirements document
- Architectural or system specification document
- Micro-architectural or RTL specification document
- Detailed algorithms document
- Design validation document
- Documentation availability standards

An effective design methodology begins with an emphasis on detailing and documenting the project requirements and specifications. *The exact format of the documents is not as important as the content of the documents.* However, for consistency, a project should specify common tools and document file formats, whether they be ASCII text or that of a word processor.

documentation distribution

In addition to actually generating project documents, it is important to define how these documents are disseminated to the team. Many teams use project web pages on company intranets to make project information available. For some projects, it is also necessary to ensure that documents are accessible from all the computer platforms that are in use on the project. For instance, some team members might be using UNIX workstations, while others might be using PCs. Thus, it may be necessary to make choices that are cross-platform compatible.

assertions and project documentation

Project documentation is also an important part of an effective assertion methodology. The project documents are a valuable resource of data that *describes where assertions are needed.* They also provide *details on what the assertions should be validating.* Requirements documents should be used to describe assertions that are placed on externally visible interfaces. Detailed design documents describe assertions needed on internal interfaces and components. Each type of project document provides additional details that can aid the creation of effective assertions.

EDA and internal tools

Tools are an integral part of a design process. EDA vendors continually provide new tools and tool enhancements. Choosing the tools that are used on a project at the *beginning* of the project enables the design team to make specific plans for each task. The tools used on a project may include a combination of the following:

- Source code management tools
- Lint checking tools (design style encouragement)
- RTL simulators
- Synthesis tools
- Back-end tools
- ATPG toolset
- Formal property checking tools
- Assertion libraries
- Productivity scripts
- Databases (test plans, coverage data, metrics)
- Debug tools

Decisions on tools are not limited to commercial tools. Rather, decisions should be made about all aspects of the design and verification process, including how to manage coverage data and organize and track test plans.

assertions and tools

Decisions about tools are also integral to your assertion methodology. Some vendor tools use proprietary assertion conventions, while others support industry standard assertion libraries (such as the OVL). Hence, project teams must carefully consider whether the assertion specification language chosen for the project is supported by the tools used on the project. This becomes an important issue if components of the design are treated as IP and delivered to other organizations or projects using a different set of tools.

RTL styles and conventions

Important, but often overlooked aspects of an effective design methodology, are RTL styles and conventions such as:

- RTL coding styles, coding rules, and coding restrictions
- RTL naming conventions
- IP reuse rules or restrictions
- Functional coverage conventions
- Linting rules

Linting tools continue to offer enhanced features and have become a standard part of design methodologies. However, other conventions can also be used to improve the success of a design. For instance, defining a set of RTL naming conventions, coding styles and rules, and coding restrictions allows designers to produce consistent code that is easily readable by all members of the team. A major reason for coding rules and restrictions is to help steer designers away from RTL code that is not sythesizable or prone to errors. Examples of general RTL coding conventions include the *Reuse Methodology Manual* [Keating and Bricaud 2002] and *Principles of Verifiable RTL Design* [Bening and Foster 2001].

With the increasing size and complexity of today's designs, reuse of IP within these designs has become more important. It is essential that conventions be defined and followed in order to improve the reusability of IP across multiple designs.

Finally, functional coverage models (see Chapter 5, "Functional Coverage"), as well as general coverage processes and coverage goals, must be defined at the beginning of the project. While coverage tools are effective in many ways, there are steps that designers can take that allow the specific coverage tools to recognize design elements (such as state machines) in an easier manner. For example, some tools use pragmas to specify state machines.

assertions and project conventions

An assertion methodology must also include consistent coding styles and conventions. Good coding practices and linting checks are just as applicable for assertions as they are for the synthesizable part of the design.

An assertion methodology convention may specify that the team will avoid using non-clocked procedural assertions due to the issues described in Chapter 4, "PLI-Based Assertions" on page 101. Another convention may be that the team will add interface assertions at the top of each module, versus the bottom of each module (see section 2.2.3, "Assertion density"). As discussed previously, an assertion methodology might specify that all assertions are included in lint checks. Refer to section 2.2.2, "Best practices" for more effective practices for your assertion methodology.

When describing IP reuse conventions, ensuring that assertions are included in the IP to validate the use of all external interfaces is a requirement. With assertions in place, the users of the IP will get immediate feedback if the IP is used incorrectly.

Assertions also provide a form of design coverage themselves and must be considered as part of the overall coverage strategy of a project. Refer to section 5.2.2, "Types of traditional coverage metrics" on page 126 for more details on coverage models.

Support infrastructure

The support infrastructure of a project is utilized throughout the project each and every day. The support component of a project relies on consistent and reliable guidelines and resources such as:

- Naming files
- Structuring design source directories
- Defining available computer resources
- Defining personnel resources

If the infrastructure has not been appropriately defined at the beginning of the project, the team may well be forced to make changes and migrate previous implementations to align with changing definitions. For instance, changing the directory structure of a Verilog model requires designers to become familiar with new file locations and initiate modifications for makefiles, filelists, and scripts.

assertions and infrastructure

An effective assertion methodology requires decisions about resources allocated to the project. In addition to engineering personnel, this includes computer resources to support the tools that use the assertion methodology. Using assertions generally requires more computing time per individual simulation run (in our experience, 10%-30% for multi-million gate designs containing thousands of assertions, yet inefficient PLI-based assertions could require 2-10x more time) than simulation efforts without them. However, people and computer resources are typically planned for on a project time-line basis. Our claim (and experience) is that the savings in debug time when using assertions actually shortens the overall project time-line (by weeks to months on some projects). Hence, identifying complex bugs sooner in the verification process, while ensuring higher quality verification, justifies the resources.

Partner coordination

Most projects draw on multiple teams or organizations to turn out a product. To ensure that all the teams work effectively together, it is important to establish communication guidelines with partners. This includes:

- Coordinating documentation deliverables with partners
- Specifying points of contact for each partner organization
- Creating formal written partner agreements where required

For instance, a company that architects a new complex computer system might require the development of three new ASICs, new circuit board assemblies, and updated software to support the new hardware. The development effort is performed by three ASIC teams, one circuit board team, and one firmware team. From this example, it should be obvious that communication between teams is important. If the three ASIC teams don't communicate, interfaces between the ASICs might be implemented differently on each chip. Also, if the ASIC teams don't communicate the configuration register settings to the software teams, the initial bring-up in the lab might be marred with incorrect initialization of the ASICs.

Defining contact persons within each team and how communication will occur between teams eliminates costly miscommunications that could possibly delay product shipment. A formal written set of expectations and deliverable may be required for some partners.

assertions and partners

Assertion specification language also plays a role in partner coordination. Models from different teams or projects are often merged into a system-level pre-silicon verification environment. From the example above, system-level verification environments which instantiate all three ASICs gain maximum benefit from assertions if all ASIC models used the same form of assertion specification. Also, the circuit board team creates a model of the circuit board that is combined with the HDL models of the ASICs in a verification environment. A similar situation arises here for any assertions written in the board models. In these cases, it is important to ensure that all models used in the system-level environment utilize compatible and consistent forms of assertion specification.

2.1.2 Design requirements

Every design begins with requirements. Sometimes these are developed in an ad hoc fashion, but ideally they are developed systematically. In essence, these requirements contain specific details on the expected behavior of the system. An example of requirement statements might be:

- "The design must process 3.0 gigabytes/sec of data"
- "The design must execute the x86 instruction set"

Capturing design requirements take several forms. This includes functionality requirements, which are usually specified first (though refined later). Silicon performance metrics are also part of the requirements; for without knowing the frequency, power, and area requirements, it is difficult to produce a good design. A third form includes requirements for delivery, and this form is important both for conveying the time required for completion and the nature of deliverables. Silicon is the common medium, but intellectual property formats are becoming common. With this soft format, the list of deliverables is much larger than delivering a functioning chip.

assertions and design requirements

Assertions play a key role in this area of a design methodology. As a project team defines requirements at any level, whether system or architectural, the team generates models of the system. Each of these elements have interface specifications describing how they should operate. Assertions should be added to the design model, even early in the process, as a method of documenting the exact intent of the specification. Depending on the method of generating these models, the assertions captured in the earlier models of the system are leveraged by the refined detailed models. For instance, a high-level system model might be generated that captures assertions for the major chip interfaces. Tools using this high-level model make use of the assertions. As more detail is added to the system-level model and the functionality of each chip is added, the original assertions are still in place to continue validating the interfaces between chips in a functional simulation environment as well as formal verification environments.

2.1.3 Design documents

Design creation requires a medium that allows for top down design, successive refinement of details, and redundancy in explanation. RTL and schematic forms do not allow for redundant descriptions (unless you include assertions as redundancy). They also suffer from the rigid requirement of a single form. Project design documents allow a mixture of forms (and detail level) to describe your design. These may include a range of documents (such as; tables, charts, block diagrams, timing diagrams, and prose). All these allow others to understand and analyze your design. In addition, one can include design analysis documentation and data from previous design iterations.

Once the initial set of design documents is complete, you are not done with the documentation effort. Without continuing to document changes as they occur, different project teams can easily get out of sync. For example, if an ASIC team changes the format of a configuration register by adding a new mode bit but fails to update the documentation, it is likely that the firmware team will program the register incorrectly. Design documents force the engineer to consider design details that may have been overlooked by starting the design without a systematic documentation process. These overlooked details can sometimes be costly and require a redesign. Generally, the need for a redesign is not found early, rather after the point at which it could have been corrected with minimal cost. Finally, when the design details have been thoroughly documented, the design is now transparent and more easily reviewed throughout a project's life.

A common set of design documents will include the following:

1 Analysis
2 Architecture
3 Preliminary design
4 Detailed design
5 Deliverable product specification

assertions and design documents

The design document provides a wealth of information that can be codified into assertions. Designs with assertions as part of the document provide additional insight into how a particular interface will operate. In this way, the assertions themselves provide additional, effective documentation.

2.1.4 Design reviews

Reviewing documentation and implementation code is crucial to the design process. This is where design teams analyze and critique design decisions for overall correctness and ensure that design performance and silicon metrics requirements are met. When the team finds incorrect structures, incomplete specifications, or unacceptable performance, the design *and* document should be updated in the problem areas and re-reviewed. In terms of time and resources, issues that are not addressed at review time become more costly to fix at a later time.

assertions and design reviews

Make assertions an integral part of the design review process. In addition to assessing design implementation, design reviews should analyze where assertions have been added. We recommend that design teams conduct reviews that specifically address adequate assertion density. (Assertion density is the number of assertions per line of code.) By including assertion review in the design review process, teams encourage designers to consciously think about their design and the corner cases and interfaces of their implementation, which are ideal locations for adding assertions.

Our experience at Hewlett-Packard Company followed this methodology, and logic design engineers found bugs in the design just by analyzing the type of assertion needed for a location within the design. The assertion was still added to the design, but this example shows that the design review process can definitely help find design bugs.

2.1.5 Design validation

make everything as simple as possible, but no simpler
-Albert Einstein

Design validation continues to be one of the dominant portions of a project cycle. Designs are more complex and larger than ever. Creative and sound methodologies are required to ensure that a design is effectively validated. EDA vendors are continuing to add features to their tool sets to increase the productivity of today's verification engineers. In addition, project teams must ensure effective verification methodologies to compliment and utilize the tool features.

assertions and design verification

Assertions are an emerging design technique that is enabling verification to be more effective. During the design creation, assertions should be at the forefront of your mind. Each time you add a new feature to the code, assess what you are writing and identify profitable assertions that will enforce correct operation and cover interesting design events. The advantages to adding assertions as you code are seen during the design validation phase of the project. A design created with an effective assertion-based methodology provides:

- a more thoroughly tested design due to an overall increase in the number of assertions (compared to adding them after coding is complete)
- added clarity of the code, which removes confusion when reviewing various code details for interactions with other features
- significantly less time spent debugging. Thinking about assertions up front can frequently clarify issues in the implementation. This removes bugs before a single simulation cycle can begin.

Refer to 1.2, "Verification techniques" on page 2 for more details on the effectiveness and benefits of assertions.

2.2 Assertion methodology for new designs

While using assertions benefits both new and existing designs, the *best results occur when design teams apply the assertion methodology from the beginning of a design process.* An effective assertion methodology plays an integral role in the debug process, and it must begin as early as possible.

For designs that are in the beginning stages of the process, refer to the discussion of the general design methodology and note the role that assertions play in each step, as described previously in section 2.1, "Design methodology". Then, proceed to the discussions that follow for a more in-depth understanding of how an assertion methodology integrates into a design project. This section includes basic *"Key learnings"* that let you know what you are up against during design verification. It offers "Best practices" to guide your process. And, it ends with ideas on both "Assertion density" and "When not to add assertions".

2.2.1 Key learnings

Buy-in from all members of the project team (from engineers to project and program managers) is essential when attempting to adopt (and fully benefit from) assertion methodologies. For instance, it is important for all members of the design team to adopt the assertion methodology to give good assertion density across the entire design. If some members of the team don't believe that using assertion is worth their time, they are not inclined to add them. Likewise, if management agrees with the importance of the assertion methodology, they will encourage all of their employees to utilize assertions. If they believe it just adds to the overall schedule, they will either not encourage the use of assertions or actually discourage their use. In both cases, the benefits of the assertion methodology will not be realized.

In this section, we draw on our previous experiences to describe design and verification challenges and important points that must be accepted by the entire team. We refer to these points as *key learnings*. As you read this section, you may think that the key learning points are just basic concepts. However, many times these key learnings are forgotten or ignored. It is through experience that these learnings are ingrained within the engineer and the team.

Our intent is to offer a reality-based understanding of what a design team is likely to face and show why it is important for the entire project team to buy into an assertion-based methodology to help solve these verification challenges.

key learning 1 **The design model is not initially correct**

This point should not come as a big revelation to any experienced engineer or manager. However, it is important to consciously acknowledge that the initial RTL model will not work directly after coding. This mind set paves the way for an effective verification process (which includes assertions) to find where the design is not correct.

This idea encourages you to select the proper design elements or architecture with careful thought and not necessarily just the first thing that pops into your mind. Critical thinking about your design points out potential problems in it. This is what a formalized debug process is about. With this learning and the information in this book, you are taking the first step towards achieving a more reliable design.

key learning 2 **Identifying and debugging design failures is difficult and time consuming**

Consider the following statement from Kernighan [1974].

> *"Debugging is at least twice as hard as writing the program in the first place."*

While many engineers also readily agree with this statement, it is not easy to state the complexity of debugging today's increasingly large designs in a way that allows you to stay on schedule and under budget. Experience has shown that assertions have a substantial positive impact on finding errors while minimizing the debug effort related to design failures.

key learning 3 **For a significantly shorter debug process, teams must accept a slightly longer RTL implementation process**

Most engineers agree that it is important to do everything possible to ensure a high quality, accurate design. However, management often gives into the pressures of schedules and budgets. In doing so, changes in the design process that increase the schedule of any portion normally are not accepted. What must be remembered in this case is that while the RTL implementation process of the project may be increased marginally by adding assertions (our studies indicate between one and three percent), the verification process is substantially reduced (up to fifty percent). (Refer to Chapter 1, "Introduction", for specific industry examples of verification successes that resulted from assertions.)

key learning 4 **Problems creep into the design during creation**

Teams must make a conscious decision to check for incorrectness and actively detect problems. Methodologies are developed to reduce the number of errors created during design capture such as through linting tools. However, it is not realistic to think that all bugs can ever be completely eliminated from the first pass design model. By recognizing this point, teams can take steps to put assertions into the design while the design is being captured. This requires that the designers recognize exactly where they are making design assumptions. These assumptions should be continuously validated during verification efforts through the use of assertions.

key learning 5 **Some portions of the design require additional verification effort**

These locations include both complicated sections of the design and intersections of blocks. Teams must make a conscious decision to seek out and record interesting combinations of events. As discussed in section 2.1.4, "Design reviews", this type of analysis improves the overall quality of the completed design. By identifying these portions of the design, assertions are fully utilized and their full benefit realized.

2.2.2 Best practices

When the key learnings discussed in the previous section are accepted (that is, buy-in is achieved from the entire project team), the following best practices will help you create your assertion-based methodology and make it effective for your design project.

use your documents

Formalize the natural language specification using assertions

As was discussed in section 2.1.3, "Design documents", the design documents and specifications are an essential resource for knowing where to add assertions. This resource can become even more beneficial if it is captured with assertions in mind. The specification is now truly verifiable, since the assertions automate the process of verifying that the implementation satisfies the specification.

when to capture assertions

Write assertions along with the RTL code

The opposite of this practice is augmenting an existing design with assertions. When assertions are written up front, bugs are often identified prior to any form of verification. However, when assertions are written at the end of design implementation, they are less effective, as many of the bugs have already been found and many of the design assumptions that assertions should be validating have been forgotten. So, you might ask why assertions should be used at all if the bugs are found without them. Assertions are not a silver bullet that will rid your design of all errors, and ensure you are exempt from respin after respin without them. Assertions improve the overall design verification effort by making it easier to find and debug failures that are found with a project's verification environments.

It has been our experience, as well as the experience of teams at many companies, that adding assertions along with the RTL code is the most effective time to add them. At this point in the project, designers are making design implementation decisions, interpreting the design requirements, and implementing the logic necessary to interface with blocks being developed by other designers. With all this information fresh on the designers mind, what better and more effective time could there be to determine where to implement assertions?

when to stop capturing assertions

Consistently implement assertions throughout the design

While it is important to add assertions as the design is beginning to be captured, the process of adding assertions should not be stopped once the initial design work is accomplished. As a general rule, more assertions constantly checking for bugs make it more likely that you will find the bugs. Additionally, as more assertions are added to the design, the assertion density of a design is improved. For these reasons, adding assertions should be a continuous process.

keep adding assertions

Analyze failures not identified by assertions to determine whether you can write new assertions that detect the problem

Adding an assertion to detect a problem that has already been found may seem counter-productive. However, there are reasons to follow this practice. First, this helps get the designer in the practice of adding assertions by identifying locations within the design where assertions are effective. Second, when you add assertions to validate known bugs, you ensure that you are able to detect another occurrence of the same design specification violation. Third, not all errors are seen by debugging a specific failure. It is often said that where you found one bug, others are sure to be close. By adding an assertion in the location of a fixed bug, additional problems may be seen by this new observation point. Adding assertions in this manner also increases your assertion density.

where to put assertions

Co-locate RTL assertions with the design code they are validating

The opposite of this practice is to place the assertions away from the code, possibly at the end of the file or in a separate file. Separating assertions from the design code they are validating removes one of the benefits that assertions provide, that of documenting the code. Additionally, adding assertions in the design code clarifies the design intent.

Including assertions with the design code also simplifies the process of adding them. Since the design file is already in an editor while the code is being created and the designer is actively thinking about the design operation, adding assertions at this point allows the designer to capture the most effective RTL assertions. Finally, it allows reviewers to easily see which portions of the code have assertions and where assertions are lacking.

reuse

Generate IP with assertions

Use assertions that are designed specifically for common design structures. Put in place methods that automatically add assertions as the common design structure is added. You can do this in a variety of ways, including using macros or libraries of modules that implement the common design structures. For instance, if your design uses FIFO structures, consider creating a library containing FIFO modules that have assertions already embedded within in them to check for underflow and overflow conditions.

assertion names

Name your assertions

All assertions should be given names or IDs. This eases the effort associated with debugging assertion condition failures. Additionally, the names of the assertions are constant as different verification tools are used (such as simulation and formal tools). Furthermore, incremental releases of the design benefit from consistent assertion names across models.

Note that while some tools can automatically generate names, others require user-specified names. However, the assertion specification form used on a project should not drive whether names are required for assertion instantiations; the assertion methodology should drive this. In all the cases mentioned above, if you use tool-generated names for assertions, the names are not easily readable and can change each time the design is built.

to use or not to use

Provide a consistent method to disable assertions

Use *ifdef* text macro capabilities with assertions to enable easy removal from a model. For SystemVerilog, the assertion constructs are directly part of the language, which means that it is unnecessary to bracket these assertions with an `ifdef. However, for any additional logic created to support the assertion that feeds into an assertion (for example, satellite FSMs to capture a special event), it is best to bracket this logic with an `ifdef construct. Example 2-1 shows an example of this concept in Verilog.

Example 2-1 Compilation control of assertions

```
`ifdef ASSERT_ON
  FIFO_check: assert @(posedge clk) (reset_n => FIFO_depth < 7);
`endif
```

If you use an assertion library, these commands should be placed within the library to reduce the designer's workload. This is already done in the OVL.

do not synthesize assertions

Provide a consistent method to prevent synthesis errors

The method for accomplishing this is tool-dependent. Many tools use comment meta-commands. These commands can wrap the assertion instantiations, much like the *ifdef* text macros capabilities. Again, depending on the tool, you may not be allowed to nest these commands. Special care should be taken to ensure the commands are used in accordance with the tool's documentation.

If you use an assertion library, these commands should be inserted in the library to reduce the designer's workload.

assertion libraries

Create libraries or templates for common assertions

Multiple designers are implementing similar assertion structures. To reduce the designer's workload, create common template libraries of assertions that provide extra logic, such as state machines that may be needed for some assertions. The OVL are an excellent example of this best practice.

design reviews

Conduct peer reviews of assertions

Peer reviews provide opportunities for designers to explore ideas for new assertions—and opportunities for designers to uncover design problems prior to the verification process. In addition, reviews identify errors within the coding of an assertion. Refer to section 2.1.4, "Design reviews" for more details about assertions and design reviews.

how effective are your assertions?

Create a process that effectively tracks identified problems

When logging bugs, you should document the technique that identified the bug. In doing so, you provide direct feedback for future projects on the effectiveness of your various verification processes. For instance, you should note whether a problem was identified with a directed or random verification environment and whether an assertion detected the bug.

One of the key learnings discussed in section 2.2.1, "Key learnings", is that while adding assertions incrementally increases the RTL development time, assertions greatly reduce the verification debug time, which can significantly improve a project's overall schedule. By capturing data on the effectiveness of assertions along with your bug tracking, you are collecting return-on-investment data that can justify the development of a new assertion-based methodology on a future project.

hook tools to assertions

Provide hooks in the verification environments to "see" assertions

Normally, when the assertion fails, the desired outcome is for the assertion to fire. However, there are some exceptions. For instance, when verifying that the design behaves in a known manner in the presence of *invalid* stimulus (for validating error correction logic), an effective assertion methodology monitors for this violation, but the assertion should not flag an error.

Since the test is known to produce an error condition, seeing the assertion fire will make the test appear as if it failed. In the best scenario, hooks allow the test to specifically "expect" the assertion to be violated. In this manner, the test will fail if the assertion is *not* violated. Alternatively, the model could be compiled with assertions disabled when you execute tests of this nature. However, this method is not highly recommended because it removes all assertions and their benefits are lost.

embedded assertion signals

Provide internal assertion "signals" to aid debugging with waveform viewers

When debugging failed simulations with the use of a waveform viewer, it is convenient to define an internal signal that is equivalent to the assertion's combinatorial expression as shown in Example 2-2. In this case, the signal `assert_valid_pnt` can be shown in the waveform viewer. This signal is inactive except when the assertion fires. This allows an engineer to quickly pinpoint the location in the waveform viewer where the assertion fired.

Example 2-2 Internal assertion signals

```
`ifdef ASSERT_ON
  assign assert_valid_pnt = (pnt >= 4'b0010) &&
                            (pnt <= 4'b1000) &&
                            (pnt != r'b0110);
  assert_always #(0, 0, 0, "illegal pointer value")
                valid_pnt (clk, reset_n, assert_valid_pnt);
`endif
```

2.2.3 Assertion density

An effective assertion methodology ensures sufficient assertion density within the RTL design. Assertion density is a measure of the number of assertions per line of code. Without sufficient assertion density, the full benefits of assertions are not realized. The goal is to have uniform assertion density with minimum holes across the entire design. Listed below are some common locations for assertions. Refer to Chapter 6, "Assertion Patterns" on page 161 for more details with examples of additional areas where assertions are effective.

general guideline

In place of RTL comments

As a general guideline, anywhere you would typically add a comment to document a potential concern, assumption, or restriction in the RTL implementation, this is an ideal location to add an assertion.

block interfaces

Block interfaces assertions

In section 2.1.2, "Design requirements", we described the benefits of adding assertions to block interfaces, particularly those that have different designers. In this case, the multiple designers identify different interpretations of a single interface's specification.

Block interfaces should have their assertions written up front when creating the architecture or specification documents for any block using the interface. With this method, you are forced to think about specific error corner cases of the interface—and check for these cases using assertions.

Do not underestimate the importance of writing assertions for block-level interfaces. For example, a logic design lead at Hewlett-Packard Company high-end server group once said, "*If a person can't write an assertion for a block or chip interface, he or she probably is not clear about the interface.*" The process of specifying assertions on block interfaces helps to clarify its correct behavior while uncovering many misconceptions (that is, bugs).

where to add interface assertions in the RTL modules

We recommend that you add all module interface assertions at the top of the RTL module. This keeps them close to the interface signals' definitions. Alternatively, you can place the interface assertions at the end of the RTL module; however, referencing the interface signal definitions becomes problematic for larger modules. Whichever location you chose for your interface assertions, we recommend that it is consistent across the entire design team.

overflow and underflow

Queue/FIFO assertions

Specify assertions to check for illegal queue or FIFO overflow and underflow conditions. In addition, assertions should monitor all design-specific corner cases of a FIFO or queue.

state machines states and transitions

State machine assertions

Specify assertions to detect invalid states and invalid transitions in a state machine. For example, in the case of a state machine with a one-hot structure, assertions monitor the state signals to ensure that no two states are ever active simultaneously.

fairness and starvation

Arbiter assertions

Specify assertions to detect fairness problems within arbiters. While fairness and starvation are difficult verification areas, assertions help reduce the associated bugs. You may need to tune the assertion equation to handle the various corner cases involved with fairness on arbiters. Keep in mind that it is better to have a few false firings of an assertion than to let a fairness bug get into silicon.

untested code

Area of code not ready for testing assertions

Often a model is released with code that is not ready for testing. Assertions should be put in these areas to quickly notify testers that features have been enabled in the verification tool that are not ready for testing.

assertion groups

Group common functionality for assertions

It is often useful to group assertions into categories to allow individual control of similar types of assertions. For instance, you may choose to group all fairness assertions together. This becomes useful if you adjust the knobs in a random test environment that stresses the system in non-realistic ways that cause such a flood of transactions that the settings of the fairness assertions fail. In this case, grouping assertions allows you to disable the arbiter class. Refer to 2.4, "Assertions and simulation" for more details on the use of assertion groups in simulations.

2.2.4 Process for adding assertions

In this section, we outline a recommended process for adding assertions to your design. We recommend that you create a set of assertions as you define block interfaces (prior to RTL coding), and then continue adding additional assertions during RTL coding. The process is thc following:

- Add assertions to each interface of a block. These assertions help to define the interface protocol, legal values, required sequencing, and so forth.
- Add assertions between interfaces of blocks. These assertions help to define how the multiple blocks interact.
- Add assertions as you code specific or unique structures within your RTL (see Chapter 6 "Assertion Patterns" and Chapter 7 "Assertion Cookbook" for examples of common structures).
- Add assertions as you code your control logic.

Following this process ensures good assertion density from the chip boundaries into the core. With this approach, incorrect behavior is isolated closer to the source of the problem. This process is also good when reviewing your RTL to ensure important assertions have not been missed

2.2.5 When not to add assertions

Use this section with caution. As a general rule, adding assertions is always a good idea. However, since there is a cost associated with adding assertions—and a cost associated with using assertions within simulation—a team should perform careful analysis prior to determining exactly where to add the assertion.

common features

Common features that are required for design operation

While this book shows that the debug and isolation benefits of using assertions is greater than the cost of adding them (and running them in simulation), the overall cost must still be recognized. As a result, every detail or aspect of a design will not warrant an assertion. The list of features where assertions should not be added include the monitoring of:

- a free running clock,
- glitch detection, asynchronous timing, or clock edge,
- assertion code that duplicates the RTL code—for example, a simple increment counter should not have an assertion that ensures the value changed by one,
- standard register D input to Q output transfers, and
- known correct components—such as a simple MUX.

duplicate checks

Design features that are validated by other methods

Some design features are checked by alternate methods—such as specific bus functional models in simulation. Also, a PCI interface may use a third party PCI protocol checker. Hence, there might not be a clear return-on-investment to duplicate the checks provided by the protocol checker with a set of assertions.

procedural assertions

Non-clocked procedural assertions

Designers should be careful with the use of non-clocked procedural assertions, as this class of assertions is prone to false firings. This is described in detail in Chapter 4, "PLI-Based Assertions" on page 101. We have found that the use of non-clocked procedural assertions is not required to obtain maximum benefit from an assertion methodology. Therefore, we recommend that you avoid them.

2.3 Assertion methodology for existing designs

An assertion-based methodology offers many of the same benefits for both new designs and existing designs. However, when implementing an assertion methodology for a mature design (for example, one that is well into the verification process), you will find fewer design problems. Hence, developing your assertion methodology at this phase of the design might not provide a dramatically clear return-on-investment.

When you use assertions with a mature design, you lose some of the benefits of capturing early designer assumptions. However, this should not keep you from using assertions, as many design assumptions can still be documented in existing designs. For example, Krolnik [1999] at Cyrix Corporation documented cases where assertions were added late in the design cycles and many design errors were unexpectedly identified for code that had been exposed to many hours of simulation. In the Cyrix case, bug reports tripled (20 issues per week rose to 60 issues per week) after assertions were added. And the time required to close out problems fell from 7 days to 2 days.

The following are best practices for existing designs that maximum the effectiveness of assertions given the limited time remaining in the project life. However, if time permits, refer to the best practices described in section 2.2, "Assertion methodology for new designs".

clarification

Use assertions to clarify understanding of the design

An existing design without assertions is missing much of the knowledge (that is, design intent) that was developed during the design process. However, important assumptions and restrictions can still be captured as assertions late in the design cycle, which will aid in future understanding (and maintenance) of the design.

use code comments

Write assertions from design code comments that imply intent

Comments such as "this will never occur" or "either of these will cause . . ." are good locations for assertions. It should be recognized, however, that comments may not have been updated when the code was modified. So use good engineering methods to determine the exact design intent.

reused components

Check properties of reused components

Components such as pass-though one-hot muxes and priority encoders that are reused throughout the design are ideal locations for assertions. When you consider the number of instantiations of these components throughout the design, you will realize that you are actually adding many assertions. If these common components are contained within a library, a little amount of work adding assertions within the library definition will impact a large portion of the design.

module interfaces

Write assertions for block interfaces

Interface protocols often have well-defined rules. Translate these rules into a set of assertions. This practice is particularly effective for providing a clear return-on-investment when reusing the block on future designs.

2.4 Assertions and simulation

A number of steps and methodology features can make your project's assertion experience much more productive, especially with regard to simulations. This section dives into features that should be a part of an effective assertion methodology. While the specific details of many of these features vary depending on the assertion specification form you use; these areas should be well defined for each project.

global enable or disable

Global enable or disable for assertions

Your assertion methodologies must provide a mechanism for enabling or disabling each assertion. We recommend that assertions be enabled by default and that you use a mechanism to disable them. This is often through the use of a global enable signal—or through an 'ifdef macro pre-processing step. You must add this mechanism to each verification environment and enable or disable it at the appropriate time.

It is also useful to separate assertions into a common group. This could be according to functionality or location. With this approach, each assertion group uses a different enable signal. This allows fine-grain control for various groups of assertions during simulation.

assertion clocks

Global clock versus local clock control

How assertions use clocks is specific to the assertion specification form you use. However, you should define a general strategy for using clocks with assertions. For example, you can use `'TOP.assert_clk` as the source clock for assertions to use a global clock. By using a global clock, you have control over the sampling of assertions to eliminate races.

assertion error reporting

Assertion error reporting facility

Implement a verification environment that permits easy management of assertion reports produced by simulation. For instance, a script that keeps volume simulations running must be able to manage multiple assertion reports, check for assertion firings when determining if a test failed, and archive the assertion reports along with other logs for failed simulations.

severity levels

Assertion severity levels

The assertion methodology you choose should support a variety of severity levels. This allows flexibility for designers as they add functional coverage. The assertion reporting facility discussed in the previous point should also support multiple severity levels. Different severity levels are used to determine when an assertion should end the simulation immediately and when it should just report the condition but continue the simulation.

quiescent state assertions

End of simulation assertions

We have found end of simulation checks extremely useful. For example, after a simulation test completes, it is critical to be aware of all outstanding transactions to determine if conditions in the design are preventing forward progress. Similarly, it is useful to know if a queue or FIFO structure contains unread data—or if there were any critical FSMs not returned to their initial state. These problems could indicate a deadlock situation. Hence, a quiescent check on a state variable, counter, or pointer at the end of simulation can uncover many hard-to-find problems.

The OVL **assert_quiescent_state** assertions is useful for performing this type of check. In addition, in Chapter 4, "PLI-Based Assertions" we demonstrate how to create a PLI routine that automatically checks a quiescent condition by performing an automatic callback at the end of simulation.

error thresholds

Alterable assertion error threshold detector

Your simulation environment should make process decisions (that is, take actions) based on the status of an assertion firing. These process decisions include: *stop, finish, print a message, continue, and increment a counter for errors.* Your assertion methodology should provide facilities to control or limit specific action based on a configured threshold. For example, how many assertion violations are required before taking a specific action? What is the limit on the number of times a unique failure should be reported? The OVL provides many examples of methodology facilities automatically built into the library.

error message requirements.

Assertion report messages

Messages reported by assertions should contain the following default information, which is used to locate the failure:

- Time of error
- Location within testbench or design hierarchy of error
- Physical location (file, line number) of error RTL code
- Severity of reported failure - error, warning, info
- Additional user-specified message and details

Composing this information into a standard message format allows for consistent extraction (for example, a *perl* script) and fast location and diagnosis of the failure. Without this complete set of information, it is difficult to isolate the exact location and time for the failure.

Even more important than this default information is the message the user contributes to the specific assertion. The user error message should contain information about the nature of the failure. This information is important to speed up the debug of the problem. The user message should contain the following information:

- What is the problem
- Where (what structure, interface)
- Optionally—who should investigate the problem

An example of an error message with this information is:

```
"Illegal command on trans_lak interface. See Jeff"
```

2.5 Assertions and formal verification

In this section, we discuss a potential role for formal property checking to play in an assertion-based verification flow. We begin with a discussion of a formal verification framework by detailing the steps required to perform formal property checking. We then outline a methodology for applying formal property checking in an industry setting.

2.5.1 Formal verification framework

In this section, our goal is to introduce the basic elements of formal property checking and in so doing, convey a sense of both its inherent power and limitations. Steps required to perform formal property checking (for example, *model checking*) include:

- compile a formal model of the design
- creating a precise and unambiguous specification
- applying an automated and efficient proof algorithm

Each of these steps are briefly discussed below.

compile a formal model

In the first step of the formal property checking process, we create a formal model of the design by compiling a synthesizable hardware description (for example, a Verilog RTL model) into a form accepted by the property checker. For the purpose of our discussion, hardware designs are finite state concurrent systems. For example, the value of the *current state* of the system can be determined at a particular point in time by examining all state-elements of the system. The *next state* of the system can be computed as a function of the system's current state value and design input values[1]. A current state—next state pair describes one particular *transition relation* of the system. For example, (s_i, s_{i+1}) is a transition relation, where s_i represents a current state of the system, and s_{i+1} represents one next state possibility directly reachable from s_i. Usually, a transition relation describes a set of all possible state transitions among states, represented in some data structure like a BDD.

A *path* at state s is an infinite sequence of states $\pi = s_0 s_1 s_2 \ldots$, which represents a forward progression of time and a succession of states. Note that a simulation trace is one example of a path. A set of paths represents the *behavior* of the system. Hence, a formal model can be created by compiling a synthesizable model of the design into as a state transition graph structure, referred to as a *Kripke structure* [Kripke 1963].

A Kripke structure M is a four tuple $M = (S, S_0, R, L)$, which consist of:

- S a finite set of *states*
- S_0 a set of *initial states*, where $S_0 \subseteq S$
- $R \subseteq S \times S$ a *transition relation*, where for every state $s \in S$, there is a state $s' \in S$, such that the state transition $(s, s') \in R$
- $L: S \rightarrow 2^{AP}$, where L is a function that labels each state with a set of atomic propositions that are true at that particular state

A Kripke structure models the design using a graph, where a node represent a state, and an edge represent transition between states.

creating a formal specification

In the next step of formal property checking, we specify properties as assertions of the design that we wish to verify. In Chapter 1, we informally defined a property as a proposition of expected design behavior (that is, *design intent*). The following is a more formal definition of a property:

Definition 2-1 ***property***: a collection of logical and temporal relationships between and among subordinate Boolean expressions, sequential expressions, and other properties that in aggregate represent a set of behavior (that is, a path). [Accellera PSL-1.0 2003]

We define a safety property as follows:

1. The next state is derived from the cone-of-logic leading into the input to a state-element. This can also be represented as a transition function $\delta(s, I)$.

Definition 2-2 ***safety property***: A property that specifies an invariant over the states in a design. The invariant is not necessarily limited to a single cycle, but it is bounded in time. Loosely speaking, a safety property claims that *something bad* does not happen. More formally, a safety property is a property for which any path violating the property has a finite prefix such that every extension of the prefix violates the property. [Accellera PSL-1.0 2003]

For example, the property, "the signals `wr_en` and `rd_en` are mutually exclusive" and "whenever signal `req` is asserted, signal `ack` is asserted within 3 cycles" are safety properties.

We define a liveness property as follows:

Definition 2-3 ***liveness property***: A property that specifies an eventuality that is unbounded in time. Loosely speaking, a liveness property claims that "something good" eventually happens. More formally, a liveness property is a property for which any finite path can be extended to a path satisfying the property.

For example, the property "whenever signal `req` is asserted, signal `ack` is asserted some time in the future" is a liveness property.

Underlying many property languages are formalism known as *propositional temporal logics,* which allows us to reason about sequences of transitions between states. Two formalisms for describing sequence propositions are *branching-time temporal logic* [Clarke and Emerson 1981][Ben-Ari et al. 1983] and *linear-time temporal* logic [Pnueli 1977]. CTL is an example of branching-time logic. The temporal operators of this formalism allow us to reason about all paths originating from a given state. Whereas in the case of LTL (a linear-time temporal logic), the temporal operators allow us to reason about events along a single computation path. In this book, we introduce the Accellera Property Specification Language (PSL) [Accellera PSL-1.0 2003]. Although PSL supports both branching-time and linear-time temporal logic. However, in Chapter 3 "Specifying RTL Properties", we focus only on the linear-time temporal component (that is, the PSL Foundation Language instead of the PSL Optional Branching Extension) since it is generally easier for the designer reason about the behavior of hardware design in terms of simulation traces.

applying a proof algorithm

Once we have created a formal model representing the design and a formal specification precisely describing a property that we wish to verify, our next step is to apply an automated proof algorithm. For example, given a formal model of a design described as a Kripke structure $M=(S, S_0, R, L)$, and a temporal logic formula f expressing some desired property of the design, the problem of proving correctness involves finding the set of all states in S that satisfy f:

$$\{s \in S \mid M, s \models f\}$$

where $s \models f$ means the property represented by temporal formula f holds at state s.

Note that the formal model satisfies the specification *if and only if* all initial states (that is, $\forall s_i \in S_0$) are in the set of all states that satisfies f (that is, $\{s \in S \mid M, s \models f\}$). Figure 2-1 graphically illustrates (at a high level) one proof algorithm used to find the set of all states in S that satisfy f.

Figure 2-1 **Fixed-point reachable states**

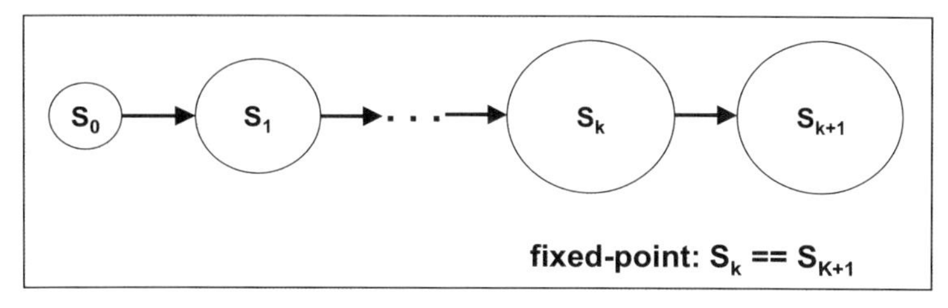

The illustrated proof algorithm we use is known as *reachability analysis* using *image computation*. The algorithm begins with a set of initial states S_0, as shown in Figure 2-1. Using the transition relation R, as previously discussed, we calculate within one step (that is, a tick of the clock) all reachable states from S_0. This calculation process is referred to as *image computation*. The new set of reachable states is S_1 in our example. We iterate on this process, generating a new set of reachable states at each step that grows monotonically, until no new reachable states can be added to the new set (that is, a *fixed-point* occurs when $S_{k+1} == S_k$).

Note that for a safety property described by the temporal formula f, we can validate that f holds on each new state calculated during the image computation step.

proof results

For this fixed-point proof algorithm, one of three different results occurs:

1 **Pass**. The process reaches a *fixed-point*, and the formula f holds on all reachable states. Hence, we are done (that is, the design is valid for this property).

2 **Fail**. The process has yet to reach a fixed-point, and the formula f was determined not to hold on a particular state s_i, which was calculated during the search. Hence, a *counter-example* (that is, a path $\pi = s_0 s_1 s_2 \ldots, s_j$) can be calculated back from the bad state s_j to an initial state $s_0 \in S_0$. This counter-example is then used to debug the problem.

3 **Undecided**. The process aborts prior to reaching a fixed-point due to a condition known as *state-explosion* (that is, there are too many states for the proof engine to represent in memory).

In the following section, we discuss a few techniques that might address the state-explosion problem.

Formal property checking tools use a number of different proof algorithms. A detailed discussion of these specific *proof algorithms, creating formal models*, and *temporal logics* is beyond the scope of this book. For in-depth discussions on these subjects, we suggest Clarke et al. [2000] and Kroph [1998].

2.5.2 Formal methodology

As formal research matures and approaches a level of sophistication required by industry (beyond the bounds of research and early adopters), we must take steps to ensure a successful transfer (scaling) to this more demanding level. One step is to fundamentally change design methodologies such that we move from ambiguous natural language forms of specification to forms that are mathematically precise and verifiable. Furthermore, these languages must lend themselves to automation. Formal *property specification* is the key ingredient in this methodological change. The end result is higher design quality through:

- *improved understanding of the design space*—resulting from the engineer's intimate analysis of the requirements, which often uncovers design deficiencies prior to RTL implementation
- *improved communication of design intent* among multiple stakeholders in the design process
- *improved verification quality* through the adoption of assertion-based verification techniques

Although the need for methodological change is clear, transitioning formal verification technology into an industry design environment has been limited by a lack of methodology guidelines for effective use.

Recall that state explosion is one of the difficulties encountered when attempting to apply formal to an industry setting. When attempting to prove correctness of assertions on an RTL implementation, a full proof is not always achievable. However, the value of functional formal verification is not limited by any means to full proofs. In reality, the value lies in finding bugs faster or earlier in the design cycle and finding difficult bugs missed by traditional simulation approaches, which in turn increases confidence in the correctness of the design while decreasing time to market.

Another difficulty encountered when attempting to apply formal to an industry setting is the methodological requirement for accurately specifying environment constraints. These are used by the formal engine to limit the exhaustive search to a valid set of legal behaviors. Note that the work used to create block-level environmental constraints for a formal engine can often be re-used as block-level interface assertions during full-chip and system simulation. Hence, there is a return on investment for specifying block-level interfaces that include, as previously state, *improved understanding of the design space, improved communication of design intent,* and *improved verification quality.*

2.5.2.1 Handling complexity

In this section, we discuss techniques typically used to handle the state explosion problem when proving properties on industrial RTL models.

Choose appropriate RTL. The first step in handling complexity is to initially choose the right level of RTL to apply formal. For example, RTL contained in control-intensive logic is better suited for formal property checking than RTL modeling data path logic. Size of the RTL component (in terms of state directly related to a property) must be considered. Other factors that influence the RTL selection are design-related. For example, not every RTL component (that is, module, block, or unit) is a good candidate for stand-alone verification. Interesting properties may require more logic to be included beyond our selected RTL component. This can be problematic since many internal interfaces are rarely documented. Furthermore, the additional logic not included with our RTL component that we wish to verify may be too complex to model as environment constraints. Nonetheless, if we choose the appropriate RTL wisely, we can have a high degree of success at formally verifying properties on RTL components.

Property decomposition. We recommend that complex sequential assertions be split into simpler assertions. For example, break a req-ack handshake down into its component elements (arcs on a timing diagram). This *think static rather than dynamic* approach works well for formal proofs.

Compositional reasoning. One technique for handling the state explosion problem is to partition a large unverifiable component into a set of smaller, independently verifiable components. This technique is referred to as *compositional reasoning*. For example, a large super-block component can be partitioned (often quite naturally) into a set of smaller block and sub-block components. When verifying a property of one of these partitioned components, you must specify a set of constraints that model the behavior of the other components (that is, the environment for the component under verification).

We define a constraint as follows:

Definition 2-4 ***constraint***: A condition (usually on the input signals) that limits the set of behavior to be considered by the formal engine. A constraint may represent real requirements on the environment in which the design is used, or it may represent artificial limitations imposed in order to partition the verification task. [Accellera PSL-1.0 2003]

Gradual semi-exhaustive verification. Although in theory, compositional reasoning using constraints sounds attractive, when applying formal property checking within an industrial setting, a more modest approach is generally used. We refer to this approach as *gradual semi-exhaustive formal verification via restrictions*. The advantage of this approach is that it has the potential of flushing out complex bugs as quickly as possible using formal verification to search a large state space.

Essentially, this approach is a gradual development of a formal verification environment around the RTL component you selected using restrictions.

This approach has the following benefits:

- Allows the user to control the state space explored to prevent state explosion using restrictions
- Enables us to initially turn off portions of the design's functionality—and then gradually turn on additional functionality as we validate the design under a set of restrictions
- Allows us to refine the constraint model into more general assumptions without initially encountering state explosion
- Provides an easier method of debugging by selecting, and thus controlling, the functionality in the environment that is enabled

We define a restriction as follows:

Definition 2-5 ***restriction***: A statement that the design is constrained by a given artificial property and a directive to verification tools to consider only paths on which the given property holds. [Accellera PSL-1.0 2003]

A restriction may reduce a set of opcodes to a smaller set of legal values to be explored during the formal search process. Or a restriction may limit the component's mode settings to read only during one phase of a proof, and then re-prove with a write mode restriction. Other examples include restricting the upper eight bits of a 16-bit bus to a constant value while letting the lower eight bits remain unconstrained during the formal search. Then, shifting the restriction to a new set of bits and re-proving with the new bus restrictions. It is important to note that even with the use of restrictions, the number of scenarios that the formal verification engine explores is very large, and complex errors will be detected under these conditions.

We demonstrate a restriction later in Section 2.5.3 "ECC example".

exhaustive proofs

The second technique used in an industrial setting, which is often used *after* the semi-exhaustive bug-hunting approach, is to relax the restrictions into general interface assumptions in an attempt to *prove* properties on the partitioned component. The advantage of performing the semi-exhaustive bug-hunting approach first using restrictions, as opposed to exhaustive proofs, is that if we cannot prove the property under the restriction, then we cannot prove it using general assumptions. Hence, we must employ other techniques (such as abstraction) if a proof is required.

We define an assumption as follows:

Definition 2-6

assumption: A statement that the design is constrained by a given property and a directive to verification tools to consider only paths on which the given property holds. [Accellera PSL-1.0 2003]

Note the subtle distinction between assumptions and restrictions related to our goal of applying formal technology in an industrial setting. For restrictions, our goal is to find bugs and clean up the partitioned components of the design using formal techniques. We are under no obligation to validate restrictions (either in simulation or formal verification). Using assumptions, however, our goal is to prove correctness—which can be a more difficult task. Often, we convert assumptions into assertions, which we then attempt to prove on neighboring components of the design. This strategy is known as *assume-guarantee reasoning* [Grumberg and Long 1994]. If an assumption is too difficult to formally prove, we use simulation to validate these assumptions as interface assertions.

2.5.2.2 Formal property checking role

identify where to apply formal

In this section, we discuss the role formal property checking plays at various phases within a design flow. The first step in the process is to identify good property candidates that provide a clear return on investment (ROI) for the effort involved in the formal verification process and likelihood for success (LFS). Examples include properties related to portions of the design that:

- have historically resulted in respins due to bugs (hence, ROI)
- are estimated to be difficult to verify (or it will be difficult to achieve high coverage) using traditional simulation means (hence, ROI)
- are contained in control-intensive logic versus data path logic (hence, LFS)
- are supported with enough bandwidth from the design team to adequately define required environment constraints when a full proof is required (hence, LFS)

identify when to apply formal

In section 1.4 "Phases of the design process" on page 14, we defined the role of assertion specification at various stages within a design flow. In this section, we discuss the role and goal of applying formal verification at various phases of design. The level of expertise required at each phase various depending on the verification goals.

Architectural verification. Formal verification has been successfully applied to proving architectural properties on shared memory consistency protocols (for example, cache coherence or sequential consistency protocols) as well as well as other architectural considerations (for example various arbitration schemes). The goal of this phase of formal verification is to flush out high-level architectural bugs prior to RTL implementation. However, successful architectural formal verification, in general, requires a verification team with a high level of expertise. In part, this expertise requirement comes from the need to create abstract models of the system that are *formal-friendly*.

Concurrent design and verification. Formal verification can be applied early during the RTL development phase in an attempt to flush out bugs prior to module integration into the system verification environment. In general, this is a low-effort task (which could be higher depending on the particular engineer's goals). As the engineer codes assertions into the RTL implementation, formal property checking combined with interface *restrictions* attempt to find bugs.

Block-level regression. Formal verification, when applied to the block-level, offers much more than a low-effort, early bug hunting tool. On the contrary, the strategy offers a means to deliver high quality blocks to the chip integration environment. Although the initial effort, before chip integration, does allow for early bug hunting, formal property checking's value extends beyond the initial stage. To provide a quick path for finding bugs and saving precious debug time during regression, it can also be performed every time the team modifies the block-level RTL code. This especially makes sense after a team makes the initial constraint investment at the block-level, which allows a formal tool to quickly prove the block-level assertions.

Targeted formal proof. Formal property checking can be applied on the set of properties previously identified as good candidates (that is, clear return on investment and likelihood for success). The effort required to perform formal property checking on an RTL model can obviously range from low (for trivial properties) to very high for a complex RTL implementation or property. Often compositional and assume-guarantee reasoning combined with some degree of abstraction are employed in an attempt to prove properties on complex designs. The effort required is mostly a function of the RTL and the property.

Post-silicon verification. We have successfully applied formal property checking during post-silicon verification. When a bug is identified in the lab, a formal test environment is created around the RTL implementation containing the bug. A property associated with the bug is created, and then the error is demonstrated on the RTL model using formal property checking combined with a formal testbench (that is, environmental properties used as constraints). Once the corrected RTL implementation is available, it is instantiated into the formal testbench and the formal property checker is used again to verify the fix. Note: Like targeted formal proofs, this can take a fair amount of effort to exhaustively prove the property on the corrected RTL.

2.5.3 ECC example

In this section, we use an error correction code (ECC) example based on Richards [2003] to demonstrate how to use *assumptions* and *restrictions* within a proof. In general, ECC algorithms are not sequential in nature. In fact, for most ECCs the proof can be formed in a single (yet complex) combinatorial step. Even though we are not demonstrating the power of a sequential search with this example, the constraint techniques we demonstrate can be applied to other more complex sequential circuits.

Error correction is the process of detecting bit errors during data transfer—and correcting these bits can be done in either software or hardware. For high data rates, error correction must be done in special-purpose hardware because software is too slow. The ECC bits are generally computed by an algorithm that is implemented as a set of exclusive OR trees in hardware. For our discussion, an *ECC generator block* computes the ECC. Each data bit contributes to more than one ECC bit. Hence, by a careful selection of data bits in the algorithm that directly contribute to the calculation for a specific ECC bit, it is not only possible to detect a single-bit error, but actually identify which bit is in error (including the ECC bits). In fact, the computed ECC is usually designed so that all single-bit errors are corrected, while all double-bit errors are detected (but not corrected).

For our discussion, an *ECC check block,* which consists of a set of exclusive OR trees, recomputes the ECC from the transmitted data bits. The output of the recomputed ECC exclusive OR network is called a *syndrome*. If the syndrome is zero, no error occurred. If the syndrome is non-zero, it can be used to index into a look-up table[2] to determine exactly which bit is in error and then correct it. For a multi-bit error, no match will occur in the lookup table.

Figure 2-2, demonstrates a formal testbench created to exhaustively verify the ECC implementation. The *ECC generator block* reads the *m* bit-wide `data_in` bus and calculates an *n* bit-wide `ecc`, which it outputs as part of an *m+n* bit-wide `data+ecc` bus. The *ECC check block* reads the `data+ecc` bus and recomputes a new ECC syndrome from the data bits, which is used for error detection and correction. For our formal testbench, we create a single level *m+n* bit-wide exclusive OR error injection circuit. This enables us to inject single bit errors during the proof.

Figure 2-2 **ECC single-bit error detect and correct proof**

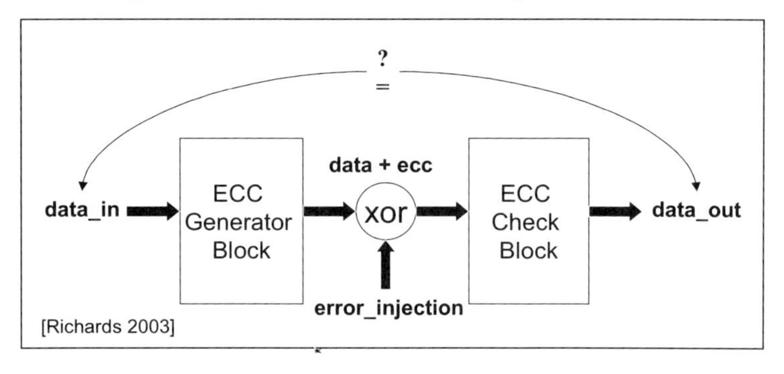

There are two techniques used to verify correct behavior of the combined ECC generator and check blocks, using assumptions and restrictions. First, for a single bit error, we can write an assertion that the *ECC check block* will always detect and correct single bit errors as shown in Example 2-3.

Example 2-3 ***PSL* assertion that the ECC detects and correct single-bit errors**

```
// constrain the error_injection to all zeroes, or a one-hot value
assume always ((error_injection & (error_injection - 1))==1'b0);

// assert that the data_in equals the data_out for single bit errors.
assert always (data_in == data_out);
```

2. The look-up table could be implemented in hardware, firmware, or even software.

The specific details for the PSL syntax are discussed in Chapter 3 "Specifying RTL Properties". However, for our example, we are using PSL to specify that the `data_in` value will always equal the `data_out` value for a single bit error. We specify a single-bit error possibility as a *zero or one-hot* assumption on the multi-bit variable `error_injection` by writing the following Verilog expression:

```
(error_inection & (error_injection - 1)) == 1'b0
```

A formal engine will use this one-hot assumption when it explores all combinations of single bit errors (that is, each bit for the `error_injection` bus will assume a one during some point in the search, effectively injecting all possible single bit error combinations into the *ECC Check block*).

Using the combined assertion and assumption, the formal proof engine exhaustively explores all possible input values and single-bit error injection combinations.

Note that if the $m+n$ bit-wide `data+ecc` and `error_injection` bus is too large, the proof could terminate with an *undecided* result. One techniquc to address this problem would be to use a combination of assumptions and restrictions. For example, we could restrict all bits with the exception of the lower four bits of the `error_injection` bus to a zero (using the PSL `restrict` construct), and then make a *zero or one-hot* assumption on the lower four bits (using the PSL `assume` construct). After proving this simpler model, we would repeat the proof by first shifting the four-bit *zero or one-hot* assumption to a new set of bits and restricting all other `error_injection` bits to zero. This process is repeated until all bits of the `error_injection` variable have had the opportunity to assume one. Ultimately, we will explore all single bit error possibilities as we shift the restrictions and assumption across the `error_injection` bits.

A similar proof can be constructed to determine if multi-bit errors are detected correctly by the *ECC check block*.

2.6 Summary

In this chapter, we focused primarily on components of an effective *assertion-based methodology* related to some of the broader design methodology considerations We then discussed how to apply an assertion-based methodology targeting *new designs*—followed by a discussion of how to apply an assertion-based methodology targeting *existing designs*. Next, we discussed simulation-based methodology considerations specifically related to assertions. Finally, we presented an overview of formal property checking and methodological considerations for applying formal in an industrial setting.

It is now up to you to choose the elements that best fit your specific project needs. Consider the concepts and guidelines we presented in this chapter when you create your project-specific assertion methodology—and then encourage your entire design team to consistently follow your methodology. By reviewing the *key learnings* with your team, you put them in a better position to fully appreciate the benefits that an assertion-based methodology provides.

CHAPTER

3

SPECIFYING RTL PROPERTIES

In this chapter, our goal is to introduce general concepts related to property specification. Then, we will apply these concepts as we introduce emerging RTL specification standards (that is, assertion libraries and languages). Initially, we compare and contrast the Accellera PSL 1.0 property specification language proposal [Accellera PSL-1.0 2003] with the Open Verification Library [Accellera OVL 2003]. We then introduce the proposed SystemVerilog 3.1 assertion constructs [Accellera SystemVerilog-3.1 2003]. Each of the assertion standards we discuss has its own merits. Our objective is to help the engineer understand the advantages (and limitations) of the various assertion forms and their usage model. This will prepare readers to select appropriate specification forms that suit their needs (or preferences).

Current users might argue in favor of the subtle advantages they perceive their favorite assertion language possesses (possibly even a different language than what we present). Ultimately, what matters is that you simply choose an assertion language and use it. When you incorporate any form of RTL assertion in your design and verification process, you will significantly improve the overall quality of your design and dramatically simplify its verification debug effort.

Some readers might skip this chapter entirely, and jump directly to the *good stuff* (that is, the examples in Chapters 6 and 7). However, at some point, these readers may find it helpful to return to this chapter to broaden their understanding of basic property language concepts and common definitions. These are covered in the beginning sections of this chapter. Our goal is to build a basic foundation of property specification and language concepts that is *useful for the RTL designer*, without delving too deeply into a theoretical discussion of automata theory.

try our assertion-based design concepts using the OVL

Many of the examples in this book use the OVL, which can be used with today's simulators. Our goal is to ensure that readers can implement and explore the various concepts we present using their existing IEEE-1364 Verilog or IEEE-1076 VHDL simulators. It is important for the reader to understand the basic assertion concepts we present in this book through working examples, such as the OVL. After mastering the general concepts we present, the reader is in position to quickly learn any property language or emerging standard for RTL assertion specification.

3.1 Definitions and concepts

Before we introduce various forms for expressing RTL assertions, it is helpful to consider definitions for two fundamentals: *property* and *event*. The reader should focus on the concepts presented in this section and not any specific syntax used to express these ideas. Details related to various assertion language syntax and semantics are discussed near the end of this chapter, as well as in the appendices.

3.1.1 Property

a property consists of a Boolean and temporal layer

Informally, a *property* is a general behavioral attribute used to characterize a design. More formally, we can define a property as:

> *A collection of logical and temporal relationships between and among subordinate Boolean expressions, sequential expressions, and other properties that in aggregate represent a set of behavior* [Accellera PSL-1.0 2003].

When studying properties, it is generally easier to view their composition as three distinct layers:

- the *Boolean layer*, which is comprised of Boolean expressions (for example, Verilog or VHDL expressions)
- the *temporal layer,* which describes the relationship of Boolean expressions over time
- the *verification layer*, which describes how to use a property during verification

Defining (or partitioning) a property in terms of the abstract layer view enables us to dissect and discuss various aspects of properties. However, you will find that it is quite simple to express design properties. Thus, the three-layer view is merely a way to explain concepts and should not convey a sense that the actual language syntax is complex.

To aid in studying property concepts, all examples in the following sections are presented using the Accellera PSL property specification language, unless otherwise noted.

Boolean layer

A property's Boolean layer is comprised of Boolean expressions composed of variables within the design model. For example, if we state that "*signal en1 and signal en2 are mutually exclusive*" (that is, a *zero-or-one-hot* condition in which only one signal can be active high at a time), then the Boolean layer description representing this property could be expressed in Verilog as shown in Example 3-1.

Example 3-1 A property's Boolean layer expressed in Verilog

```
!(en1 & en2) // enables are mutually exclusive
```

time ambiguity

Notice that we have not associated any time relationship to the statement: "*signal en1 and signal en2 are mutually exclusive*". In fact, the statement by itself is ambiguous. Is this statement *true* only at time 0 (as many formal tools infer), or is it *true* for all time?

Temporal layer

together, the Boolean and temporal layers form the basis of a property

A property's temporal layer permits us to describe the Boolean expressions' relationships to each other over time. Thus, all time ambiguities associated with a property are removed. For example, if *signal en1 and signal en2 are always mutually exclusive* (that is, for all time), then a temporal operator could be added to the Boolean expression to state precisely this. Temporal operators allow us to specify precisely when the Boolean expression must hold.[1] Example 3-2 demonstrates this point using the PSL temporal operator *always* combined with a Verilog Boolean expression.[2]

Example 3-2 A property's temporal layer expressed in PSL

```
always !(en1 & en2) // enables are always mutually exclusive
```

1. The term *hold* in this context means that the design exhibits behavior described by a specific Boolean expression, when the Boolean expression evaluates *true*.

2. Note that a PSL *property* definition does not end with a semicolon (;), while *assertions* (which are built on top of properties) do end in a semicolon.

We discuss additional PSL temporal operators used to specify relationships of multiple Boolean expressions over time in detail later in this chapter. And, we provide examples throughout the remainder of the book.

Verification layer

the verification layer for a property defines how to use it during verification

While a property's Boolean and temporal layers describe general behavior, they do not state how the property should be used during verification. In other words, should the property be asserted, and thus checked? Or, should the property be assumed as a constraint? Or, should the property be used to specify an event used to gather functional coverage information? Hence, the third layer of a property, which is the *verification layer*, states how the property is to be used.

Consider the following definitions for an assertion and a constraint.

- *Assertion* - A given property that is expected to hold within a specific design. The PSL assert directive would be associated with the property to specify an assertion.
- *Constraint* - A condition (usually on the input signals) which limits the set of behavior to be considered during verification. A constraint may represent real requirements (e.g., clocking requirements) on the environment in which the design is used, or it may represent artificial limitations (e.g., mode settings) imposed in order to partition the verification. In this case, the PSL assume or restrict directives would be associated with the property to specify a constraint.

Constraints and assertions are described in further detail in Section 3.2.2.

Look again at the property *signal en1 and signal en2 are mutually exclusive*. Example 3-3 shows this property with the PSL assert directive. This states that the property is to be treated as an assertion during verification.

Example 3-3 Specifying a PSL property as an *assertion* for verification

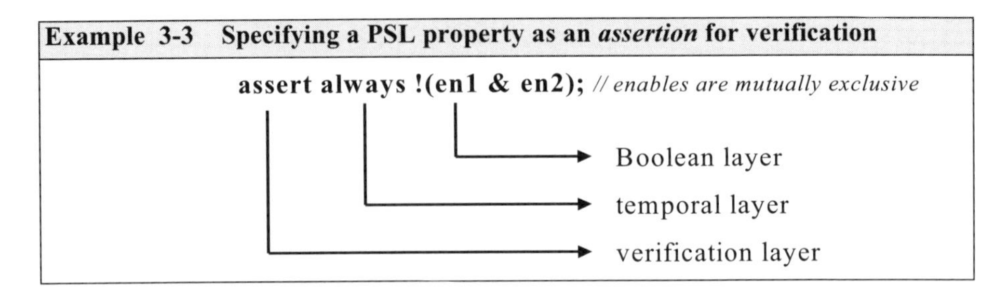

3.1.2 Events

When discussing design properties in the context of verification (and in particular, simulation), it is helpful to understand the concept of a verification event. An *event* is any user-specified property that is satisfied at a specific point in time during the course of verification. A *Boolean event* occurs when a Boolean expression evaluates *true* in relation to a specified sample clock. A *sequential event* is satisfied at the end of a sequence of Boolean events.

In Example 3-4, if the sequence `c_mem_access` followed by `c_write` is satisfied during simulation (for example, at time unit 100), then we can claim that an event has occurred in our verification environment at that specific point in time. However, if the event is never satisfied, then our verification test was unable to verify some key aspect or functionality of our design. In other words, our testing and input stimulus was insufficient.

Example 3-4 A PSL functional coverage point

```
cover {c_mem_access; c_write};
```

The PSL cover directive permits the designer to designate the property as a *functional coverage point*. Chapter 5, "Functional Coverage" discusses additional aspects of creating functional coverage models through property specification.

3.2 Property classification

Properties are often classified in the context of their *temporal* and *verification* layers. Furthermore, properties can be classified by their *evaluation* method (that is, concurrent or sequential activation). This section describes the various classifications of properties.

3.2.1 Safety versus liveness

As previously defined, a property is a general behavioral attribute that is used to characterize a design. It is generally expressed in a format that enables us to reason about sequences of Boolean expressions over time. Hence, a property is often classified by its temporal layer. This section defines the two property classifications that are based on the temporal layer: *safety* and *liveness*.

invariant property

A *safety* property is also known as an invariant; which informally states that, for all time, nothing bad should happen. Thus, it is a property that must evaluate to *true* for all sample points of time. The sample point could be defined by either an explicit clock associated with the property or an inferred clock.

Figure 3-1 illustrates a safety property. The Boolean expression q in this example must always evaluate to *true* at every clock cycle.

Figure 3-1 **invariant property**

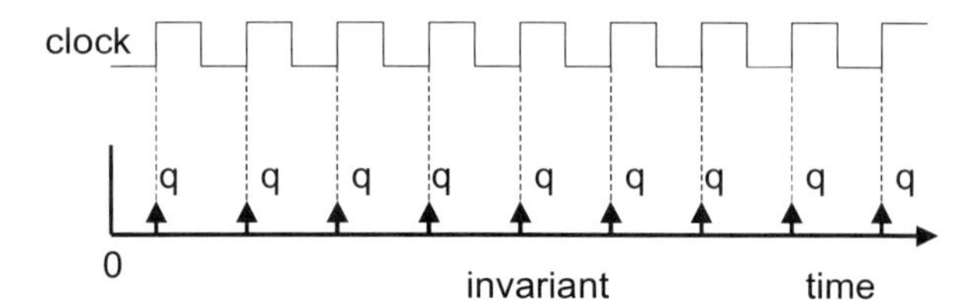

The arrows along the time axis represents the sampling of the Boolean expression q at every positive edge of signal clock (for this example). For an safety property, the sampled Boolean expression must always evaluate to *true*, which is represented by the up arrow.

liveness property

A liveness property is a property that specifies an eventuality that is unbounded in time. Loosely speaking, a *liveness* property claims that something good *eventually* happens. For example, the property "*whenever signal* `req` *is asserted, signal* `ack` *is asserted sometime in the future*" is a liveness property.

3.2.2 Constraint versus assertion

classifying properties as constraints or assertions

In addition to safety and liveness, a property can be classified according to its verification layer as either a *constraint* or an *assertion*.

One example of a *constraint* is a property that specifies the range of allowable values (or sequences of values) permitted on an input. The design cannot be guaranteed to operate correctly if its input value (or sequence of values) violates a specified constraint.

Alternatively, a property that describes that the expected design output behavior must remain valid or *true*, is an example of an *assertion*. For any permissible sequence of input values applied to a design (which means that all input constraints are satisfied), all assertions will evaluate to *true* if the design is functioning correctly.

The functions of constraints and assertions are dependent on the verification tool and environment. During *simulation*, both constraints and assertions can be treated as monitors (that is, dynamic property checkers) that check for compliance. During *formal verification*, constraints bound the static formal search engine to the design's legal input space, while assertions are treated as targets (that is, properties that must be proved) for formal analysis.

3.2.3 Declarative versus procedural

A *declarative property* describes the expected behavior of the design independent of its RTL procedural details. Hence, it is not necessary to understand the procedural code to understand the required expected behavior. On the other hand, a *procedural property* describes the expected behavior of the design in the current context (or frame of reference) at a particular line within the procedural code. Hence, it is necessary to understand the details of the procedural code to fully understand the expected behavior. Expressing interface properties declaratively is generally more natural than expressing these properties procedurally, since interface requirements are typically independent of the details of the block's implementation. However, capturing internal RTL implementation's design intent procedurally generally reduces the amount of extra code required to express these properties (particularly if the assertion is deeply nested with case and if statements).

concurrent versus sequential

A design model typically consists of a static, hierarchical structure, in which primitive elements interact through a network of interconnections. The primitives may be built-in simple functions (for example, gates) or larger, more complex procedural or algorithmic descriptions (for example, VHDL processes or Verilog always procedural blocks). Within a procedural description, statements execute in sequence. However, within the design as a whole, the primitives and communication interact concurrently.

Just as the design model itself involves a collection of concurrent elements (represented in a *declarative* fashion) and sequential elements (represented in a *procedural* fashion), properties may also be represented either as declarative or procedural. Hence, a *declarative assertion* is a statement (outside of a procedural context) that is always active and is evaluated concurrently with other layers or primitives in the design. A *procedural assertion*, on the other hand, is a statement within the context of a process (for example, a Verilog procedural block) that executes sequentially, in its turn, as the procedural code executes.

3.3 RTL assertion specification techniques

The system architect and verification engineer are instrumental in specifying global design assertions that must be verified (that is, independent of lower-level implementation details). However, quite often the verification engineer lacks sufficient in-depth knowledge of the implementation details to provide effective white-box assertion density. On the other hand, during the course of RTL development, the design engineer makes many lower-level assumptions about the design's environment as well as other implementation assumptions. Experience has shown that if design assumptions or concerns are not captured during the process of RTL implementation, then many lower-level implementation decisions, details, and properties are lost (that is, they will not be verified).

In this section, we demonstrate various forms for RT-level assertion specification. This includes the OVL [Accellera OVL 2003], PSL formal property language [Accellera PSL-1.0 2003], and SystemVerilog 3.1 proposed assertion constructs [Accellera SystemVerilog-3.1 2003]. Note that a more in depth discussion of these proposed standards is presented in Appendices A, B, and C.

3.3.1 RTL invariant assertions

The most basic form of RTL assertions is simple invariant (safety) properties, as discussed in Section 3.2.1. Examples of safety properties in RTL code include:

- it is never possible to overflow a specific FIFO
- it is never possible to read and write to the same memory address simultaneously
- it is never possible to generate an address out of range

In this section, we present examples of both an OVL and a PSL invariant assertion.

OVL invariant

Example 3-6 demonstrates the coding of invariant assertions directly into the RTL using the OVL for a simple FIFO circuit. This example is based on the UART 16550 core designed by Jacob and Mohor [2001]. The RTL assertion specifies that the blocks interfacing with the FIFO should never overflow or underflow the FIFO buffer. In other words, attempting to perform a *push* operation when the FIFO is full will result in an assertion violation. Similarly, it is a violation to perform a *pop* operation when the FIFO is empty.

OVL assert_never invariant

The Accellera OVL assert_never monitor, demonstrated in Example 3-6, accepts three arguments:

1. a clock expression
2. a reset expression
3. a user-specified Boolean expression

The Boolean expression is evaluated on every rising edge of the sample clock (when the reset signal is not active), and the monitor asserts that the Verilog Boolean expression will never evaluate to *true*. If the overflow assertion in Example 3-6 is violated, the following (default) error message is logged during simulation:

```
OVL_ERROR : ASSERT_NEVER : VIOLATION : : severity 0 : time 105 :top.my_FIFO.no_overflow
```

In addition to checking that a Verilog Boolean expression *never* evaluates to *true*, OVL provides the assert_always monitor to check that a Boolean expression *always* holds. For additional OVL feature details, such as customizing error message or severity levels, as well as an overview of additional monitors contained within the library, see Appendix A.

PSL invariant

specifying safety properties with PSL

A formal property language offers an alternative to instantiating assertion monitors directly in the RTL source (as demonstrated in Example 3-6). The expected behavior could be specified using a formal property language, such as the Accellera PSL formal property language. We could express the same overflow and underflow assertions using PSL, as demonstrated below in the Verilog Example 3-5.

Example 3-5 ***PSL* overflow and underflow assertion**

```
assert never (reset_n && {push,pop}==2'b10 &&
cnt==FIFO_depth)   @(posedge clk);

assert never (reset_n && pop && cnt==0)
  @(posedge clk);
```

Example 3-6 Verilog FIFO overflow and underflow assertion example

```
module FIFO (data_out, data_in,
             clk,  FIFO_clr_n, FIFO_reset_n,
             push, // push strobe, active high
             pop   // pop strobe, active high
            );
// FIFO parameters
   parameter FIFO_width  = `FIFO_WIDTH;
   parameter FIFO_depth  = `FIFO_DEPTH;
   parameter FIFO_pntr_w = `FIFO_PNTR_W;
   parameter FIFO_cntr_w = `FIFO_CNTR_W;
   output [FIFO_width-1:0] data_out;
   input  [FIFO_width-1:0] data_in;
   input                   clk, FIFO_clr_n,
                           FIFO_reset_n,push,pop;
// FIFO buffer declaration
   reg [FIFO_width-1:0] FIFO[FIFO_depth-1:0];
// FIFO pointers
   reg [FIFO_pntr_w-1:0] top; // top
   reg [FIFO_pntr_w-1:0] btm; // bottom
   reg [FIFO_cntr_w-1:0] cnt; // count

'ifdef ASSERT_ON
   wire reset_n = FIFO_reset_n & FIFO_clr_n;

// OVL assert that the FIFO cannot overflow
   assert_never no_overflow (clk, reset_n,
       ({push,pop}==2'b10 && cnt==FIFO_depth-1));

// OVL assert that the FIFO cannot underflow
   assert_never no_underflow (clk, reset_n,
       (pop && cnt==0));
'endif

   always @(posedge clk or negedge FIFO_clr_n)
// Clear FIFO content and reset control
    if (!FIFO_clr_n) begin
      top <= 0;
      btm <= 0;
      cnt <= 0;
      for (i=0; i<FIFO_depth; i=i+1)
         FIFO[i] <= 0;
    end
// reset FIFO control
    else if (!FIFO_reset_n) begin
      top <= 0;
      btm <= 0;
      cnt <= 0;
    end
  else
```

Example 3-6 Verilog FIFO overflow and underflow assertion example

```
    case ({push, pop})
      2'b10 : // WRITE
        begin // assertion checks for overflow
          FIFO[top] <= data_in;
          top <= top + 1;
          cnt <= cnt + 1;
        end
      2'b01 : //READ
        begin // assertion checks for underflow
          btm <= btm + 1;
          cnt <= cnt - 1;
        end
      2'b11 : // WRITE & READ
        begin
          FIFO[top] <= data_in;
          btm <= btm + 1;
          top <= top + 1;
        end
      endcase
// end always

  assign data_out = FIFO[btm];

endmodule
```

Note that the PSL assertion is a declarative form of specification, which is independent of any hardware description language. In other words, at the Boolean layer PSL supports Verilog, VHDL or technically any HDL. The basic syntax for expressing a PSL never assertion (related to a Boolean expression) is shown in Example 3-7.[3]

Example 3-7 *PSL* overflow and underflow assertion

```
assert never <Boolean expression> [@<clock expression>];
```

Note that in Example 3-5, we demonstrated this assertion using a Verilog Boolean expression. For VHDL code, the appropriate VHDL Boolean expression syntax would be used.

In addition to checking that a Boolean expression never holds, PSL provides the means for checking that a Boolean expression *always* holds with the assert always keywords. For additional details, see Appendix B.

3. For a more in-depth description of the PSL never operator, particularly related to specifying sequences and properties, see Appendix B.

3.3.2 Declaring properties with PSL

As previously stated in this chapter, a *property* specifies a behavior of the design. Once defined, a property can be used in verification as an *assertion* (a property that must be checked), a *functional coverage* specification (a property the must evaluate to true during verification), or a *constraint* (a property that limits the verification input space).

PSL allows you to define named property declarations with optional arguments, which facilitates property reuse. These parameterized properties can then be instantiated in multiple places in your design with unique argument values. A property can be referenced by its name.

For example, we could specify that a and b are mutually exclusive whenever a reset_n is not active as follows:

Example 3-8 ***PSL* property declaration example**

```
property mutex (boolean clk, reset, a, b) =
    always !(a & b ) @(posedge clk) abort !reset_n;
```

reset condition

The abort clause allows you to specify a reset condition. If the abort Boolean expression becomes true at any time during the evaluation of the assertion expression, then the property holds regardless of the assertion expression evaluation.

In Example 3-9, we now create a PSL assertion for a design property where write_en cannot occur at the same time as a read_en:

Example 3-9 ***PSL* assertion for mutex write_en & read_en**

```
property mutex (boolean clk, reset, a, b) =
    always !(a & b ) @(posedge clk) abort !reset_n;

assert mutex(clk_a, master_rst_n, write_en, read_en);
```

3.3.3 RTL cycle related assertions

In this section, we demonstrate how to express assertions involving multiple Boolean expressions using the PSL next temporal operator and the OVL assert_next monitor.

PSL next operator

The OVL assert_always and assert_never monitors, and the PSL always and never temporal operators, allow us to specify an invariant (that is, a condition that must hold or must not hold for all cycles). Additional OVL monitors and PSL operators allow us to be more specific about specifying cycle timing relationships. For instance, the PSL next operator allows us to specify the cycle relationship between consecutive events (that is, the relationship between Boolean or temporal expressions). Thus, the PSL assertion in Example 3-10 states that whenever the signal `req` holds, then the signal `ack` must hold on the next cycle.

Example 3-10 *PSL* next operator

```
assert always (req -> next ack) @(posedge clk);
```

PSL Boolean implication

Note the use of the PSL Boolean implication operator `->`. In math, the implication operator consist of an *antecedent* that implies a *consequence* (for example, `A -> C,` which reads `A` implies `C`). If the antecedent is true, then the consequence must be true for the implication to pass. If the antecedent is false, then the implication passes regardless of the value of the consequence.

Continuing our example, if the `ack` is expected to hold on the 3rd cycle after the `req`, then the assertion would have to be coded in a more complicated form, as shown in Example 3-11.

Example 3-11 *PSL* multiple next

```
assert always (req -> next (next (next ack))) @(posedge clk);
```

PSL next repetition operator

Although the specification for multiple next cycles shown above is valid, PSL provides a more succinct mechanism that utilizes the repetition operator [i], where *i* is a constant value. As shown in Figure 3-2, next[3] states that the operand is required to hold at the 3rd next cycle (rather than at the very next cycle).

Figure 3-2 **assert always (req -> next[3] ack) @(posedge clk);**

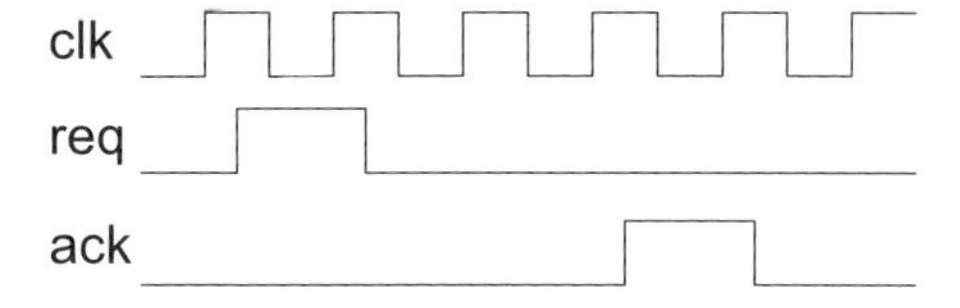

OVL assert_next

The Accellera OVL assert_next monitor has semantics that are similar to the PSL next operators, as shown in the following PSL assertion.[4]

4. The PSL abort operator specifies a condition that removes any obligation for a property to hold. The left operand of the abort operator is the property to be aborted. The right operand of the abort operator is the Boolean condition that causes the abort to occur.

Example 3-12 *PSL* abort operator

```
assert always (req -> next ack) @(posedge clk) abort !reset_n;
```

However, this same example could be coded in the RTL by instantiating a Verilog OVL assert_next monitor as shown in Example 3-13.

Example 3-13 *OVL* assert_next

```
assert_next my_req_ack (clk, reset_n, req, ack);
```

The OVL equivalent of the PSL assertion demonstrated in Figure 3-2 is demonstrated below in Example 3-14. Recall that this specifies that an ack must occur exactly three cycles after a req.

Example 3-14 *OVL* assert_next with 3 *number of clocks* parameter

```
assert_next #(0,3) my_req_ack (clk, reset_n, req, ack);
```

In Example 3-14, the #(0,3) parameters represent the *severity level* (0) and *number of clocks* (3) required for the sequence. A severity level of 0 is the highest severity, which will cause simulation to halt. And, the ack signal must be satisfied three clocks after req, as specified by the *number of clocks* parameter. For additional details on the OVL assert_next parameter options, see Appendix A.

3.3.4 PSL and default clock declaration

PSL provides a means for specifying a *default clock* expression, which enables you to define a property or sequence without explicitly specifying a clock. For example, we could re-write Example 3-12 using a default clock declaration as shown in Example 3-15:

Example 3-15 *PSL* default clock

```
default clock = (posedge clk);
assert always (req -> next ack) abort !reset_n;
```

For additional details on the default clock syntax, see Appendix B.

For simplicity, we have coded many of the PSL examples throughout the book without an explicit clock. For these examples, you can assume that a default clock was previously defined—just like we did in Example 3-15.

3.3.5 Specifying sequences

In this section, we discuss specifying sequences with PSL and OVL. First, we explore the power of PSL to support a concise coding style. Then, we demonstrate how the OVL is implemented to check sequences.

Checking sequences with PSL

sequences of Boolean expressions

The basic PSL temporal operators described in the previous sections (that is, always, never, and next) can be combined to create complicated assertions. However, writing such assertions is sometimes cumbersome, and reading (and understanding) complicated assertions can be equally difficult.

The assertion shown in Example 3-16 states that the following sequence must occur:

- if signal `req` is asserted
- and then in the next cycle, signal `ack` is asserted
- and then in the following cycle signal `halt` is not asserted
- then, starting at that cycle, signal `grant` is asserted for two consecutive cycles

Example 3-16 *PSL* **sequence specified with the next operator**

```
assert always
  (req -> next(ack -> next(!halt -> (grant & next grant))));
```

Figure 3-3 Sequence.

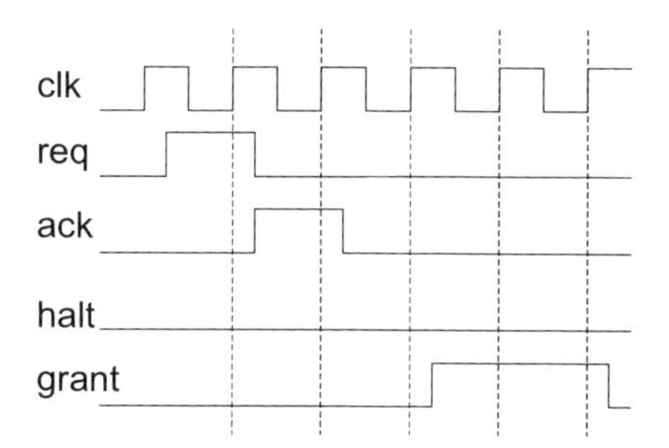

sugar extended regular expressions (SEREs)

PSL provides an alternative way to reason about sequences of Boolean expression that is more concise and easier to read and write. It is based on an extension of regular expressions, called *sugar extended regular expressions*, or SEREs.

SEREs describe series of Boolean events by specifying a sequence for which each Boolean expression in the series must hold over contiguous cycles. A rudimentary SERE is a single Boolean expression describing a Boolean event at a single cycle of time. More complex sequences of Boolean expressions can be constructed using the SERE concatenation operator (;). Example 3-17 shows the specification that three Verilog Boolean expressions `A&B`, `C|D`, and `E^F` must hold consecutively.

Example 3-17 ***PSL* SERE - sequence of Boolean expressions**

```
{A&B; C|D; E^F}
```

The sequence is matched if the following three assertions hold:

- on the first cycle, the Boolean expression `A&B` holds
- on the second cycle, the Boolean expression `C|D` holds
- on the last cycle, the Boolean expression `E^F` holds

Often, an implication operator is used to start the sequence. Thus, if our specification states: *if signal* `req` *is asserted, then in the next cycle, signal* `ack` *must be asserted, and in the following cycle, signal* `halt` *must not be asserted,* then the property could be written using a SERE as shown in Example 3-18.

Example 3-18 ***PSL* SERE for req, ack, !halt sequence**

```
assert always req -> next {ack; !halt};
```

Note that if the req signal does not hold, then the sequences (defined by the SERE) that start in the cycle immediately after req are not required to hold.

SERE [*] and [+] consecutive repetition operators

Repetitions within sequences. Like regular expressions found in most scripting languages (such as Perl or TCL), PSL allows the user to specify repetitions when specifying sequences of Boolean expressions (SEREs). For instance, the SERE *consecutive repetition operator* `[*m:n]` describes repeated consecutive concatenation of the same Boolean expression (or SERE) that is expected to hold between m and n cycles, where m and n are constants.

If neither of the range values are defined (that is, `[*]`) then the SERE is allowed to hold any number of cycles, including zero. Hence, the empty sequence is allowed. Also note that the repetition operator [+] is shorthand for the repletion [*1:inf], where the inf keyword means infinity.

For example, the SERE {a; b[*]; c[3:5]; d[+]; e} describes the following sequences:

- the Boolean variable a holds on the first cycle of the sequence
- and then, on the following cycle, there must be zero or more b's that must hold
- this is followed by three to five c's that must hold
- which is followed by one or more d's that must hold
- which is finally followed by e that must hold.

Sequence implication. In Example 3-18, we demonstrated a simple Boolean implication, in which a Boolean expression implied a sequence. Often, it is desirable for the completion of one sequence (that is, a prerequisite sequence) to imply either a property or another sequence. Hence, the *suffix implication* family of PSL operators enables us to specify this type of behavior.

The PSL *suffix implication* operator |-> can be read as: *If the left hand side prerequisite sequence holds, then the right hand side sequence (or property) must hold.* The | character symbolizes the completion of the prefix sequence, which is then followed by the implication operator ->, implying the suffix sequence.

Let us reconsider Example 3-16. Suppose that the two cycles of grant should start the cycle after !halt. We could code this as shown in Example 3-19.

Example 3-19 ***PSL*** **suffix implication**

```
assert always {req; ack; !halt} |-> {1; grant, grant};
```

Or, we can simplify the code by using the repetition operator as shown in Example 3-20.

Example 3-20 ***PSL*** **repetition operator**

```
assert always {req; ack; !halt} |-> {1; grant[2]};
```

Note that the last Boolean expression in the prerequisite sequence overlaps (occurs at the same time as) the first Boolean expression in the suffix sequence. In other words, the !halt Boolean expression in the prerequisite SERE overlaps with the first item in the suffix SERE. Hence, we add the 1 (*true*) Boolean expression for this overlap, which moves time forward by one cycle. An alternative way to code Example 3-19 is shown in Example 3-20.

Example 3-21 ***PSL* suffix implication with next operator**

```
assert always {req; ack; !halt} |-> next {grant[2]};
```

However, PSL provides a simpler way to do this using the *suffix next implication* operator |=>. The |=> operator takes us forward in time by one clock cycle, which permits us to specify the property in Example 3-19 as shown in Example 3-22.

Example 3-22 ***PSL* suffix next implication**

```
assert always {req; ack; !halt} |=> {grant[2]};
```

Declaring sequences within PSL

In PSL, sequences can be declared and then reused with optional parameters.

For example, we could define a *request-acknowledge* sequence with parameters that allow redefining the `req` and `ack` variables as follows:

Example 3-23 ***PSL* sequence declaration**

```
sequence req_ack (req, ack) = {req; [*0:2]; ack};
```

Once defined, a sequence can be reused and referenced by name within various PSL properties.

Sequence operators within PSL

PSL provides a number of sequence operators, useful for composing sequences. In this section we introduce the *sequence fusion* operator and the *sequence length matching AND* operation. The additional operators are described in Appendix B.

Sequence fusion (:). The *sequence fusion* operator (:), constructs a SERE in which two sequences overlap by one cycle. That is, the second sequence starts at the cycle in which the first sequence completes. For example, to specify that an active `grant` overlaps with the last active `req` in a set of sequences of one, two, three, or four consecutive `req` signals, we would code:

Example 3-24 *PSL* `grant` overlapping very last `req`

```
{req[*1:4}:{grant}
```

Sequence length matching AND (&&). The *sequence length-matching AND* operator (&&) constructs a SERE, in which two sequences both hold at the current cycle, and furthermore both complete in the same cycle. An example of the sequence length matching AND operator is demonstrated in Example 3-51 PSL PCI basic read transaction on page 98.

Checking sequences with the OVL

The OVL assert_cycle_sequence monitor is a useful way to check a contiguous sequence of Boolean expressions. In the Verilog OVL, the sequence `A`, followed by `B`, followed by `C`, followed by `D`, would be expressed as a concatenation operation `{A,B,C,D}`. This expression is then passed on as the *event_sequence* argument to the assert_cycle_sequence monitor. Note that the monitor can be configured (as an option) to check either of the analogous PSL properties shown in Example 3-25 and Example 3-26.

Example 3-25 *PSL* Boolean expression implies sequence

```
always {A} |=> {B,C,D}
```

Example 3-25 states that the occurrence of the Boolean expression `A` implies the sequence `{B,C,D}` starting on the next cycle, or

Example 3-26 PSL Sequence implies Boolean expression

```
always {A,B,C} |=> {D}
```

Example 3-26 states that the occurrence of sequence `{A,B,C}` implies the Boolean expression `D` on the next cycle (see Appendix B for details).

Another example, Example 3-27, asserts that when a `'WRITE` cycle starts, which is then followed by a `'WAIT` statement, then the next opcode must have the value `'READ`.

Example 3-27 ***OVL* assert_cycle_sequence**

```
assert_cycle_sequence #(0,3) init_test (clk,1,
        {r_opcode == 'WRITE,
         r_opcode == 'WAIT,
         r_opcode == 'READ});
```

This is analogous to the PSL assertions shown in Example 3-28.

Example 3-28 ***PSL* opcode sequence**

```
assert always
  (r_opcode=='WRITE) -> next {r_opcode=='WAIT; r_opcode=='READ};
```

3.3.6 Specifying eventualities

PSL eventually! operator

In section 3.3.3, "RTL cycle related assertions", we introduced the next operator, which allows us to move forward exactly one cycle. However, we might not want to explicitly specify (or we may not know) the exact timing relationship between multiple events. Hence, the PSL eventually operator allows us to move forward without specifying exactly when to stop. The assertion that *whenever a* `req` *is asserted, then an* `ack` *will eventually be asserted* would be coded in PSL as shown in Example 3-29.

Example 3-29 ***PSL* eventually operator**

```
assert always req -> eventually ack;
```

PSL until operator

The PSL until operator provides another way to reason about a future point in time, while specifying a requirement on a Boolean expression that must hold for the current cycles moving forward (that is, until a *terminating property* holds). See Example 3-30.

Example 3-30 ***PSL* until operator**

```
assert always req -> next (!req until ack);
```

This assertion states that whenever signal `req` is asserted, then starting at the next cycle, signal `req` will be de-asserted until signal `ack` is asserted. For this example, Boolean expression (that is, signal) `ack` is the terminating property.

up to but not necessarily including

The until operator is a *non-inclusive* operator; that is, it specifies the left operand holds up to, but not necessarily including, the cycle where the right operand terminating property holds. As such, the sub-property `(!req until ack)` specifies that `req` will be de-asserted up to, but not including, the cycle where `ack` is asserted. Thus, if signal `ack` is asserted immediately after the cycle in which the signal `req` is asserted, then the de-assertion of req is not required.

up to and including

Alternatively, the until_ operator is an *inclusive* operator; that is, it specifies the left operand holds up to and including the cycle where the right operand terminating property holds. Thus, if the req signal is required to be de-asserted (that is, `!req`) at least one cycle after the initial `req`, then until_ would be used to specify this property as shown in Example 3-31.

Example 3-31 *PSL* event bounded window pattern

```
assert always req -> next(!req until_ ack);
```

Example 3-31 states that whenever signal `req` is asserted, then `!req` will be asserted during the next cycle (whether or not `ack` is asserted), and it will remain asserted through (and including) the cycle where `ack` is asserted.

weak versus strong operators

One additional note concerning the PSL eventually, until and until_ operators: these are known as *weak* operators. A *weak* operator makes no requirements about the terminating condition, while a *strong* operator requires that the terminating condition eventually occur. For example, the `ack` signal in Example 3-31 is not required occur prior to the end of verification (for example, at the end of simulation) for the weak until_ operator. The eventually!, until!, and until_! are all strong operators. For additional details concerning *strong* (!) and *weak* operators, see Appendix B.

OVL event bounded window checkers

OVL assert_window

The OVL contains a set of *event bounded* window checkers that permit us to specify an eventuality class of assertions. This allows an assertion similar to Example 3-30 to be captured in the RTL using an OVL assert_window monitor, as shown in Example 3-32.

Example 3-32 *OVL* event bounded window

```
assert_window req_ack (clk, reset_n, req, !req, ack);
```

For additional details concerning the assert_window, as well as other OVL event bounded (and time bounded) window checkers, see Appendix A.

3.3.7 PSL built-in functions

PSL contains a number of built-in functions that are useful for modeling complex behavior. In this section, we describe the prev(), rose(), and fell(), which are used throughout various example in the book.

- prev (bit_vector_expr [, number_of_ticks])

  returns the previous value of the `bit_vector_expr`. The `number_of_ticks` argument specifies the number clock ticks used to retrieve the previous value of `bit_vector_expr`. If `number_of_ticks` is not specified, then it defaults to one.

- rose (boolean_expr)

  The built-in function rose() is similar to the posedge event control in Verilog. It takes a Boolean signal as an argument and returns a true if the argument's value is 1 at the current cycle and 0 at the previous cycle, with respect to the clock of its context, otherwise it is false.

- fell (boolean_expr)

  The built-in function fell() is similar to negedge in Verilog. It takes a Boolean signal as an argument and returns a true if the argument's value is 0 at the current cycle and 1 at the previous cycle, with respect the clock of its context, otherwise it is false.

3.4 Pragma-based assertions

A number of proprietary, vendor- and tool-specific approaches have been developed in recent years that provide designers the ability to embed verification assertions in their RTL design code. Because of a lack of standardization within this area, many of these approaches must rely on text-macros [Bening and Foster 2001], or meta-comment mechanisms to specify design assertions. These are followed by a pre-processing step that attempts to model the assertion semantics in HDL (or via PLI) in a way that is transparent to the user. One attractive feature with either the text-macro or meta-comment approach is that they have no side effect on existing tools that read the RTL code (that is, they can be ignored by existing tools as comments). Furthermore, this approach permits the designer to embed new assertion languages within an existing HDL standard.

The Accellera PSL formal property language permits us to specify assertions of the design independent of the implementation code; that is, the HDL code. However, it is often desirable to capture assertions directly in the HDL source code during RTL implementation. To achieve inter-operability, many companies have coordinated an effort for embedding assertions within RTL code using the common pragma approach. Example 3-33 illustrates a technique that embeds PSL assertions directly within Verilog code.

Example 3-33 ***PSL* embedded pragma**

```
// PSL assert always req -> next ack;
```

For PSL properties and assertions spanning multiple lines, Example 3-34 demonstrates a recommended technique.[5]

Example 3-34 ***PSL* embedded pragma spanning multiple lines**

```
/* PSL
  property req_ack = always req -> next ack;
  assert req_ack;
*/
```

It is encouraging to see that multiple EDA vendors are working together to ensure inter-operability for PSL embedded assertions. However, we recommend that you consult your specific tool's documentation prior to embedding a PSL assertion into your RTL code. Please note that the syntax for a PSL property and assertion is fixed through the Accellera standard. However, the embedding mechanism across multiple design languages (for example, a pragma syntax) is outside the scope of the PSL formal property language.

3.5 SystemVerilog assertions

The SystemVerilog 3.1 proposal [Accellera SystemVerilog-3.1 2003] recommends extensions to the IEEE-1364 Verilog language that permit the user to specify assertions declaratively (that is, outside of any procedural context) or directly embedded within

5. Note in this example we declared a PSL *named property* (req_ack) and then *assert* the property in a separate statement.

procedural code.[6] In addition, SystemVerilog supports two forms of assertion specification: *immediate* and *concurrent*.

In this section, we focus on a small set of common SystemVerilog operators that we use in examples throughout the book. Details for all the SystemVerilog operators are covered in Appendix C, "SystemVerilog Assertions" on page 327.

immediate event-based semantics

Immediate assertions evaluate using simulation *event-based semantics*, similar to other procedural block statements in Verilog. There is a danger of semantic inconsistency between the evaluation of immediate assertions in simulation versus formal property checkers, since formal tools generally evaluate assertions using *cycle-based semantics* (that is, sampled off of a clock or signals) versus simulation event-based semantics. In addition, there is a risk of false firing associated with immediate assertions in simulation, which is discussed later.

concurrent cycle-based semantics

Concurrent assertions are based on clock semantics and use sampled values of variables (note, this is similar to the OVL clock semantics). All timing glitches (real or artificial due to delay modeling and transient behavior within the simulator) are abstracted away.

For a detailed discussion of SystemVerilog scheduling and semantics related to assertion evaluation, we recommend Moorby et al. [2003].

3.5.1 Immediate assertions

use immediate assertions with caution

Immediate assertions (also referred to as *continuous invariant assertions*) derive their name from the way they are evaluated in simulation. In a procedural context, the test of the assertion expression is evaluated *immediately*, instead of waiting until a sample clock occurs. When the variables in the assertion expression change values in the same simulation time slot, due to transient scheduling of events within a zero-delay simulation model, a *false firing* may occur if standard Verilog event scheduling is used. To prevent this class of false firings, evaluation of the assertion expression must wait until all potential value changes on the variables have completed (that is, the transient behavior of events in the simulation has reached a steady state). Hence, SystemVerilog 3.1 has proposed a new region within the simulation scheduler's time slot called the *observe region*, which evaluates after the non-blocking assignment (NBA)

6. Our discussion of SystemVerilog is based on the Accellera committee work as of April 7, 2003. We recommend you reference the final LRM at www.accellera.org for any subtle last minute changes in syntax. The final LRM is targeted for release in May 2003.

region—ensuring that assertion expression variable values have reached a steady state [Moorby et al. 2003]. Note that this is similar to performing a PLI *read-only synchronization* callback to get to the end of the time slot region for safe evaluation. However, there is still a potential for false firings across multiple simulation time slots with immediate assertions, often due to a testbench driving stimulus into the DUV and delay modeling. For a detailed discussion of this problem in the context of PLI routines, see section 4.1.4 "False firing across multiple time slots" on page 114.

The syntax for the SystemVerilog immediate assertion is defined as follows:

Syntax 3-1 *SystemVerilog* immediate assertions

```
// See Appendix C for additional details

immediate_assert_statement ::=assert ( expression ) action_block
action_block        ::=  statement   [ else statement_or_null ]
                       | [statement_or_null] else statement_or_null
statement_or_null ::=statement   |  ';'
```

Note that the SystemVerilog assert statement is similar to a Verilog if statement. For example, if the assertion expression evaluates to true, then an optional *pass statement* is executed. If the pass statement is omitted, then no action is taken when the assertion expression evaluates to true. Alternatively, if the assertion expression evaluates to 1'bx, 1'bz, or 0, then the assertion fails and the optional else *fail statement* is executed. If the optional fail statement is omitted, then a default error message is printed for whenever the assertion expression evaluates to false.

naming assertions

The optional assertion label (identifier and colon) associates a name with an assertion statement. And it can be displayed using the %m format code.

severity levels

SystemVerilog has created a new set of system tasks (also referred to as severity task) that are similar to the Verilog $display system task. These new tasks convey the severity level associated with an assertion's `action_block` while printing any user-defined message. The new severity tasks are:

- $fatal, which reports a Run-time Fatal severity level and terminates the simulation with an error code.
- $error, which reports a Run-time Error condition and does not terminate the simulation. Note that if the optional fail state is omitted, the `$error` is the default severity level.
- $warning, which reports a Run-time Warning severity level and can be suppressed in a tool-specific manner.
- $info, which reports any general assertion information, carries no specific severity, and can be used to capture functional coverage information during runtime.

The details and syntax for these system tasks are described in Appendix C.

Example 3-35 demonstrates a SystemVerilog *immediate* assertion for our previous FIFO example:

Example 3-35 *SystemVerilog* queue underflow check

```
always @ (push or pop or cnt or reset_n)
  if (reset_n)
    if ({push, pop}==2'b01)
      underflow_check: assert (cnt!=0) else
                          $error("underflow error at %m");
```

3.5.2 Concurrent assertions

SystemVerilog concurrent assertions describe behavior that spans time. The evaluation model is based on a clock such that a concurrent assertion is evaluated only at the occurrence of a clock tick. SystemVerilog 3.1 has proposed a new region within the simulation scheduler's time slot, called the *preponed region*, which evaluates at the beginning of a simulation time slot. Hence, the values of variables used in the concurrent assertion expression are sampled at the start of a simulation time slot and then the concurrent assertion is evaluated using the preponed sampled values in the time slot observe region. Further details on concurrent assertion sampling are described in Moorby et al. [2003].

3.5.2.1 Property declaration

reusing properties

As previously stated in this chapter, a *property* specifies a behavior of the design. Once defined, a property can be used in verification as an *assertion* (a property that must be checked), a *functional coverage* specification (a property the must occur during verification), or a *constraint* (a property that limits the verification input space).

SystemVerilog allows you to define named property declarations with optional arguments, which facilitates property reuse. These parameterized properties can then be instantiated in multiple places in your design with unique argument values. A property can be referenced by its name. A hierarchical name can be used consistent with the System Verilog naming conventions.

For example, we could specify that a and b are mutually exclusive whenever a reset_n is not active as follows:

Example 3-36 ***SystemVerilog* property declaration example**

```
property mutex (clk, reset_n, a, b);
    @(posedge clk) disable iff (reset_n) (!(a & b ));
endproperty
```

reset condition

The disable iff clause allows you to specify asynchronous resets. If the disable Boolean expression becomes true at any time during the evaluation of the assertion expression, then the property holds regardless of the assertion expression evaluation. SystemVerilog also supports the specification of properties which must never hold, using the not construct. Effectively, the not construct negates the property expression. For example, we re-code the previous example as follows:

Example 3-37 ***SystemVerilog* property declaration example with not**

```
property mutex_with_not (clk, reset_n, a, b);
    @(posedge clk) disable iff (reset_n) not (a & b);
endproperty
```

See Appendix C for specific details on SystemVerilog property syntax.

3.5.2.2 Verifying concurrent properties

After declaring a property, a verification directive assert or cover can be used to state how the property is to be used. The SystemVerilog verification directives include:

- assert—which specifies that the property is to be used as an assertion (that is, a property whose failure is reported during verification).
- cover—which specifies that the property is to be used as a functional coverage specification (that is, a property whose occurrence is reported during verification).

In Example 3-36, we now create a concurrent assertion for a design property where write_en cannot occur at the same time as a read_en:

Example 3-38 *SystemVerilog* **assertion for mutex write_en & read_en**

```
property mutex (clk, reset_n, a, b);
    @(posedge clk) disable iff (reset_n) (!(a & b));
endproperty

assert_mutex: assert property (mutex(clk_a, master_rst_n, write_en,
read_en));
```

Example 3-39 demonstrates an alternative form of directly specifying the same SystemVerilog concurrent assertion:

Example 3-39 *SystemVerilog* **simple concurrent assertions**

```
assert_mutex: assert property @(posedge clk_a)
    disable iff (master_reset_n) (!(write_n & read_en));
```

Note that a concurrent assertion may be used directly within procedural code, or alternatively stand alone as a declarative assertion within a module (that is, outside of procedural code).

See Appendix C for specific details on SystemVerilog assert and cover syntax.

3.5.2.3 SystemVerilog sequences

sequences of Boolean expressions

A *sequence* is a finite series of Boolean events, where each expression represents a linear progression of time. Thus, a sequence describes a specific behavior. A SystemVerilog *sequence expression,* like the PSL SERE previously discussed, describes sequences using *regular expressions.* This enables us to concisely specify a range of possibilities for when a Boolean expression must hold.

Example 3-40 shows how we use SystemVerilog to concisely describe the sequence "*a request is followed three cycles later by an acknowledge*".

Example 3-40 *SystemVerilog* **sequence expression with fixed delay**

```
req ##3 ack
```

specifying cycle delays

In SystemVerilog, the ## construct is referred to as a *cycle delay* operator. The number after the ## construct represents the cycle in which the right-hand side Boolean event must occur with respect to the left-hand Boolean event. For the case ##0, both the left- and right-hand Boolean events overlap in time (that is, they occur in parallel).

We can specify a time window with a cycle delay operation and a range. Example 3-41 uses SystemVerilog to describe the sequence "*a request is followed by an acknowledge within two to three cycles*".

Example 3-41 ***SystemVerilog* sequence expression with a range of delays**

```
req ##[2:3] ack
```

The previous examples are referred to as *binary delays* (that is, a delay between two Boolean expressions). SystemVerilog also permits us to specify *unary delays* (that is, a Boolean expression that begins with a delay). Examples of unary delays are as follows:

Example 3-42 ***SystemVerilog* unary delays relationship to binary delays**

```
(##0 start)      // that is, (start)
(##1 start)      // that is, (1'b1 ##1 start)
(##[1:2] start) // that is, (1'b1 ##1 start) or (1'b1 ##2 start)
```

Note that unary delays are useful when associated with implication. For example, if we want to describe a sequence in which an req must be followed by a ack within two to three cycle, which is then followed by a gnt, we would write it as follows (using the SystemVerilog implication operator |->):

Example 3-43 ***SystemVerilog* unary delays relationship to binary delays**

```
req |-> ##[2:3] ack ##3 gnt
```

Sequence declaration

In SystemVerilog, sequences can be declared and then reused with optional parameters, as shown in Syntax 3-2.

Syntax 3-2 ***SystemVerilog* sequence**

```
// See Appendix C for additional details

sequence_declaration ::=
  sequence sequence_identifier [ sequence_formal_list ] ;
      { sequence_decl_item }
      sequence_spec ;
  endsequence [ : sequence_identifier ]

sequence_formal_list ::=
  ( formal_list_item { , formal_list_item } )

sequence_decl_item ::=
    variable_declaration
  | sequence_declaration

sequence_spec ::=
  [ cycle_delay_range ] sequence_element
    { cycle_delay_range sequence_element }

cycle_delay_range ::=
    ## constant_expression
  | ## [ const_range_expression ]
```

You can replace expression names within the sequence expression via parameters specified through the sequence_formal_list. This enables us to declare sequences and reuse them in multiple properties. For example, we could define a request-acknowledge sequence with parameters that allow redefining the `req` and `ack` variables as follows:

Example 3-44 ***SystemVerilog* sequence declaration**

```
sequence req_ack (req, del, ack);
  (req ##[1:3] ack); // ack occurs within 1 to 3 cycles after req
endsequence;
```

Sequence operations

SystemVerilog defines a number of operations that can be performed on sequences, such as:

- specifying repetitions
- specifying the occurrence of two parallel sequences
- specifying optional sequence paths (for example, split transactions)
- specifying conditions within a sequence (such as the occurrence of a sequence within another sequence or that a Boolean expression must hold throughout a sequence)

- specifying a first match of possible multiple matches of a sequence
- detecting an endpoint for a sequence
- specifying a conditional sequence through implication
- manipulation data within a sequence

In this section, we focus on a small set of common SystemVerilog sequence operators that we use in examples throughout the book. Details for all the SystemVerilog sequence operators are covered in Appendix C, "SystemVerilog Assertions" on page 327.

Repetition operators

SystemVerilog allows the user to specify repetitions when defining sequences of Boolean expressions. The repetition counts can be specified as either a range of constants or a single constant expressions.

Like PSL, SystemVerilog supports three different types of repetition operators, as described in the following section.

construct repeated consecutive concatenation of sequences

Consecutive repetition. The consecutive repetition operator [*n:m] describes a sequence (or Boolean expression) that is consecutively repeated with one cycle delay between the repetitions. Note that this is exactly like the PSL [*m:n] operator. For example,

```
expr[*2]
```

specifies that `expr` is to be repeated exactly `2` times. This is the same as specifying:

```
expr ##1 expr
```

In addition to specifying a single repeat count for a repetition, SystemVerilog permits specifying a range of possibilities for a repetition.

rules for repeat counts

SystemVerilog repeat count rules are summarized as follows:

- Each repeat count specifies a *minimum* and *maximum* number of occurrences. For example, [*n:m], where n is the minimum, `m` is the maximum and `n <= m`.
- The repeat count [*n] is the same as [*n:n].
- Sequences as a whole cannot be empty.
- If `n` is 0, then there must be either a prefix, or a post fix term within the sequence specification.
- The keyword **$** can be used as a maximum value within a repeat count to indicate the end of simulation. For formal verification tools, **$** is interpreted as infinity (for example,

[*1:$] describes a repetition of one to infinity). Note that this is similar to the PSL 1.0 inf keyword.

constructs repeated (possibly non-consecutive) concatenation of a sequences

Nonconsecutive count repetitions. The nonconsecutive count repetition operator [*=n:m] describes a sequence where one or more cycle delays are possible between the repetitions. The resulting sequence may proceed beyond the last Boolean expression occurrence in the repetition. Note that this is exactly like the PSL [=m:n] operator. For example,

```
a ##1 b [*= 1] ##1 c
```

is equivalent to the sequence:

```
a ##1 !b [*0:$] ##1 b ##1 !b [*0:$] ##1 c
```

In other words, there can be any number of cycles between a and c as long as there is one b. In addition, there can be any number of cycles between a and the occurrence of b, and any number of cycles between b and the occurrence of c (that is, b is not required to proceed c by exactly one cycle).

Note, the same sequence in PSL 1.0 would be coded as:

```
{a; b[=1]; c}
```

Nonconsecutive exact repetitions. The nonconsecutive exact repetition operator [*->n:m] (also known as the *goto repetition* operator) describes a sequence where a Boolean expression is repeated with one or more cycle delays between the repetitions and the resulting sequence terminates at the last Boolean expression occurrence in the repetition. Note that this is exactly like the PSL 1.0 [->m:n] goto operator. For example,

```
a ##1 b [*-> 1] ##1 c
```

constructs repeated (possibly non-consecutive) concatenation of a Boolean expression, such that the Boolean expression holds on the last cycle

is equivalent to the sequence:

```
a ##1 !b [*0:$] ##1 b ##1 c
```

In other words, there can be any number of cycles between a and c as long as there is one b. In addition, b is required to proceed c by exactly one cycle.

Note, the same sequence in PSL 1.0 would be coded as:

```
{a; b[->1]; c}
```

First match operator

The SystemVerilog first_match operator matches only the first occurrence of possibly multiple occurrences of a sequence expression. This allows you to discard all subsequent matches from consideration.

The syntax for the SystemVerilog first match operator is described as follows:

Syntax 3-3 ***SystemVerilog*** **first match operator**

```
// See Appendix C for additional details

sequence_expr ::=
   first_match ( sequence_spec )
```

Consider an example with a variable delay specification as shown in Example 3-45.

Example 3-45 ***SystemVerilog*** **first match for req ack sequence**

```
sequence seq_1;
  req ##[2:4]ack;
endsequence

sequence seq_2;
  first_match(req ##[2:4]ack);
endsequence
```

Each attempt of sequence `seq_1` can result in matches for up to four following sequences:

```
req ##2 ack
req ##3 ack
req ##4 ack
```

However, sequence `seq_2` can result in a match for only one of the above four sequences. Whichever of the above three sequences matches first becomes the result of sequence `seq_2`. Notice that this is useful if the `ack` signal is held high for multiple cycles. The first_match prevents multiple unwanted matches from occurring.

Throughout operators

SystemVerilog provides a means for specifying that a specific Boolean condition (that is, an invariant) must hold throughout a sequence using the following construct:

Syntax 3-4 ***SystemVerilog*** **throughout operator**

```
// See Appendix C for additional details

sequence_expr::=
   boolean_expr throughout sequence_element
```

For example, to specify sequence such that an `interrupt` must not occur during an `req-ack-gnt` transaction, we would code the following:

Example 3-46 ***SystemVerilog*** **sequence with Boolean condition**

```
!interrupt throughout (req ##[2:4] ack #[1:2] gnt)
```

Dynamic variables within sequences

SystemVerilog *dynamic variables* are local variables with respect to a sampling point within a sequence. The advantage of dynamic variables (over global variables) is that each time the sequence is entered, a new local variable is dynamically created. This ensures the sampling of data in overlapping sequence is correctly related to the appropriate sequence evaluation.

In Example 3-47 we demonstrate the usefulness of dynamic variable when validating the correct input/output data relationship in a pipeline register of depth sixteen.

Example 3-47 ***SystemVerilog*** **dynamic variable to validate pipeline latency**

```
// pipeline regster of depth 16
sequence pipe_operation;
  int x;
  write_en,(x = data_in)) |-> ##16 (data_out == x);
endsequence
```

Restriction on dynamic variable usage, as well as syntax details, are defined in Appendix C.

3.5.2.4 SystemVerilog implication operators

The SystemVerilog implication operator supports sequence implication using the following constructs:

Syntax 3-5 ***SystemVerilog*** **implication operators**

```
// See Appendix C for additional details

property_implication ::=
     sequence_spec |-> [not] sequence_spec
   | sequence_spec |=> [not] sequence_spec
```

SystemVerilog provides two forms of implication: *overlapped* using operator `|->`, and *non-overlapped* using operator `|=>`. The *overlapped implication operator* `|->` is similar to the PSL *suffix implication operator* `|->` which can be read as: *If the left hand side prerequisite sequence holds, then the right hand side sequence must hold.* Likewise, the *non-overlapped implication operator* `|=>` is similar to the PSL *suffix next implication operator* `|=>`, which takes us forward in time by a single clock. For example, the non-overlapped implication operator:

```
(a |=> b)
```

is the same as the overlapped implication operator with a unary delay of one:

```
(a |-> ##1 b).
```

The following points should be noted for sequential implication.

- If the *antecedent sequence* (left hand operand) does not succeed, implication succeeds vacuously by returning true.
- For each successful match of the antecedent sequence, the *consequence sequence* (right hand operand) is separately evaluated, beginning at the end point of the matched *antecedent sequence*.
- All matches of *antecedent sequence* require a match of the *consequence sequence*.
- Nesting of implication is not allowed

3.5.3 System functions

SystemVerilog provides a number of new system functions useful when defining assertions, which include:

- **$past** (*bit_vector_expr* [, *number_of_ticks*]) returns a previous value of the *bit_vector_expr*. The *number_of_ticks* argument specifies the number clock ticks used to retrieve the previous value of *bit_vector_expr*. If *number_of_ticks* is not specified, then it defaults to one.

- **$inset** (<expression>, <expression> {, <expression> }) returns true if the first expression is equal to at least one of the subsequent expression arguments.
- **$insetz** (<expression>,<expression> {, <expression> }) returns true if the first expression is equal to at least one other expression argument. Comparison is performed using casez semantics, so ‘z’ or ‘?’ bits are treated as don’t-cares.
- **$isunknown** (<expression>) returns true if any bit of the expression is ‘x’. This is equivalent to

  ^<expression> === ’bx.

- **$countones** (<expression>) returns a count that represents the number of bits in the expression set to one. The ‘x’ and ‘z’ value of a bit is not counted towards the number of ones.

See Appendix C for additional details related to SystemVerilog.

3.6 PCI property specification example

In this section, our goal is to demonstrate a process of translating a set of natural language requirements into a set of properties. We have chosen examples from the Peripheral Component Interconnect (PCI) specification [PCI-2.2 1998]. Please note that it is not our intention to fully specify all functional requirements of the PCI—we leave this as an exercise for the reader.

You will note that many of the properties we specify in this section are at a transaction-level. Protocol specification and verification at a transaction level is more efficient than at a signal interaction level. Transaction level specification not only permits more efficient test stimulus generation—it also enables debugging and measuring functional coverage at a higher level of abstraction. Nonetheless, specifying transaction level properties is generally not efficient for formal verification (see section 2.5.2 "Formal methodology" on page 48).

Transactions are conveniently constructed by partitioning the behavior definition into a set of sequence specifications, with each sequence representing a specific behavior segment of a transaction. These sequences are then combined to form a more complex bus transaction specification. We recommend that interface protocol or transaction specification occur prior to coding the RTL, at the *specify* or *design/architect* phases described in section 1.4 "Phases of the design process" on page 14.

3.6.1 PCI overview

The PCI local bus is an industry standard, high performance 32- or 64-bit local bus architecture with multiplexed address and data lines. The bus was defined with the primary goal of establishing an industry standard high performance, low cost interconnect mechanism between highly integrated peripheral controller components, peripheral add-in boards, and processor/memory systems.

We begin our discussion of creating a PCI formal specification by illustrating the bus interface required pin list as shown in Figure 3-4. This is followed by a brief description for each required PCI signal. Finally, we demonstrate how to convert a natural language specification of the PCI bus protocol into a set of assertions.

Figure 3-4 **PCI 2.2 Required Pin List**

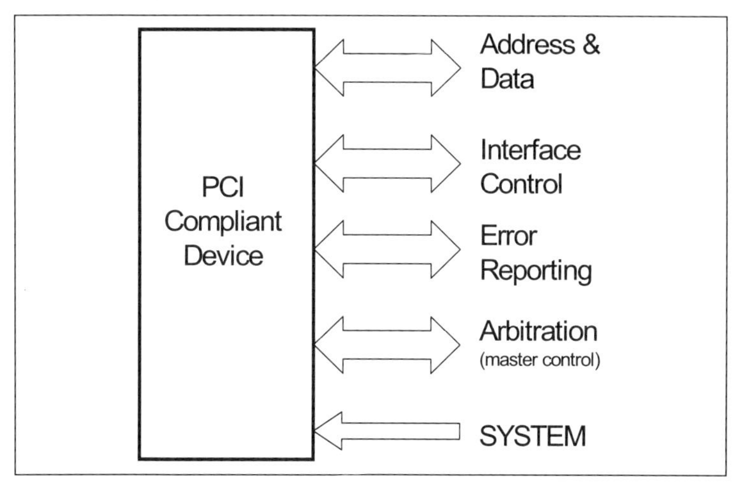

Table 3-1 **Address & Data**

Pin	Description	Direction
AD[31:0] *(ad)*	Address and Data are multiplexed on the same PCI pins.	*bi-directional*
C/BE[3:0 *(cbe_n)*	Bus Command and Byte Enables are multiplexed on the same PCI pins.	*bi-directional*
PAR *(par)*	Parity is even parity across AD[31:0] and C/BE[3:0]#.	*bi-directional*

Table 3-2 **Interface Control**

Pin	Description	Direction
FRAME# *(frame_n)*	FRAME# is activated by the current master to indicate the beginning and duration of a transaction. When FRAME# is deasserted, the transaction is in the final data phase or has completed.	*bi-directional*
IRDY# *(irdy_n)*	Initiator Ready indicates the bus master's ability to complete the current data phase of the transaction. IRDY# is used in conjunction with TRDY#. Note that a data phase is completed on any clock when both IRDY# and TRDY# are asserted.	*bi-directional*
TRDY# *(trdy_n)*	Target Ready indicates the target-selected device's ability to complete the current data phase of the transaction.	*bi-directional*
STOP# *(stop_n)*	STOP# indicates that the current target is requesting the master to stop the current transaction.	*bi-directional*
IDSEL *(idsel)*	Initialization Device Select is used as a chip select during configuration read and write transactions.	*input*
DEVSEL# *(devsel_n)*	Device Select, when actively driven, indicates the driving device has decoded its address as the target of the current access.	*bi-directional*

Table 3-3 **Error Reporting**

Pin	Description	Direction
PERR# *(perr_n)*	Parity Error is only for reporting data parity errors during all PCI transactions (except a Special Cycle not discussed here).	*bi-directional*
SERR# *(serr_n)*	System Error is for reporting address parity errors, data parity errors on the Special Cycle command, or any other system error where the result will be catastrophic.	*bi-directional*

Table 3-4 **Arbitration**

Pin	Description	Direction
REQ# *(req_n)*	Request indicates to the arbiter that this agent desires use of the bus.	*output*
GNT# *(gnt_n)*	Grant indicates to the agent that access to the bus has been granted.	*input*

Table 3-5 System

Pin	Description	Direction
CLK *(clk)*	Clock provides timing for all transactions on PCI and is an input to every PCI device. All other PCI signals, except RST#, INTA#, INTB#, INTC#, and INTD#, are sampled on the rising edge of CLK and all other timing parameters are defined with respect to this edge.	*input*
RST# *(req_n)*	Reset is used to bring PCI-specific registers, sequencers, and signals to a consistent state.	*input*

A PCI bus transaction consists of an *address phase* followed by one or more *data phases*. During the address phase, the `C/BE[3:0]#` bus command indicates the type of transaction. During the data phase, `C/BE[3:0]#` are used as Byte Enables.

Note that the # symbol at the end of the signal name indicates an active low signal. For our examples, we convert the # symbol into "_n" as part of the name to indicate an active low signal.

3.6.1.1 PCI master reset requirement

In this section, we demonstrate how to translate a simple PCI reset requirement, stated in section 2.2.1 (page 9) of the PCI Local Bus Specification [PCI-2.2 1998]. The PCI reset requirement is stated as follows:

> *To prevent AD, C/BE#, and PAR signals from floating during reset, the central resource may drive the RST# lines during reset (bus parking) but only to a logic low level; they may not be driven high.*

This is an example of a *conditional expression pattern*, described in section 6.4.1 on page 179. In Example 3-48, we have written a PSL assertion to check that the `AD`, `C/BE#`, and `PAR` signals are never driven high during reset.

Example 3-48 ***PSL* master reset assertion**

```
assert always (rst_n==0) -> !(|{ad, cbe_n, par}) @(posedge clk);
```

Note that we have used the Verilog *reduction or* operator to determine if any bit in this example is a logical one. The same assertion could be specified using a Verilog OVL implication monitor as shown in Example 3-49.

Example 3-49 ***OVL* master reset assertion**

```
assert_always master_reset (clk, !rst_n, !(|{ad, cbe_n, par});
```

3.6.1.2 PCI burst order encoding requirement

The memory address space for the PCI is defined by the bits `AD[31:2]`. the lower two bits (that is, `AD[1:0]`) are encoded to indicate the order in which the master is requesting the data transfer, as defined in section 3.2.2.2 (page 29) of the PCI Local Bus Specification. Table 3-6 specifies the legal burst order encoding for memory transactions. Hence, address bit `AD[0]` must never be set to an active high value for a memory transaction burst order request.

Table 3-6 **Burst Order Encoding**

AD[1]	AD[0]	Burst Order
0	0	Linear Increment
0	1	Reserved
1	0	Cache Wrap Mode
1	1	Reserved

Example 3-50 ***PSL* PCI legal memory transaction burst order encoding**

```
'define mem_cmd ((cbe_n == 'MEM_READ) || \
                 (cbe_n == `MEM_WRITE) || \
                 (cbe_n == `MEM_RD_MULTIP) || \
                 (cbd_n == `MEM_RD_LINE) || \
                 (cbd_n == `MEM_WR_AND_INV))

sequence SERE_MEM_ADDR_PHASE = {frame_n; !frame_n && mem_cmd};

property PCI_VALID_MEM_BURST_ENCODING =
  always {SERE_MEM_ADDR_PHASE} |-> {!ad[0]}
    abort !rst_n @(posedge clk);

assert PCI_VALID_MEM_BURST_ENCODING;
```

In Example 3-50, we code a PSL assertion to validate a correct memory burst order request. Note that this assertion uses a sequence to define a memory address phase sequence (that is, a falling edge of `FRAME#`, along with the decoding of a memory transaction from the bus command `C/BE#`). Whenever this prefix sequence occurs, then bit `AD[0]` must always be active low for a valid burst order encoding.

Note that Example 3-50 is an example of a *sequence implication pattern,* as described in section 6.4.2 on page 181. The PCI memory address phase is described by defining the sequence SERE_MEM_ADDR_PHASE, which matches sequences containing a falling edge of `FRAME#` combined with a decoding of a memory command. This forms a prefix sequence, which implies that the reserved `AD[0]` is not active high. For additional details on sequence and the suffix implication operator |->, see Appendix B.

3.6.1.3 PCI basic read transaction

In this section, we demonstrate (via an simplified example) another transaction-level property, which we construct by partitioning the transaction into a set of partial behaviors specified as sequences. A PCI basic read operation consists of the following phases:

- an *address phase*, which for a basic read consists of a single address transfer in one clock
- a *data phase*, which includes one transfer state plus zero or more wait states

The address phase occurs on the first clock cycle in which `FRAME#` is asserted. For a basic read transaction, there must be at least one turn around cycle between the address phase and the data phase. A data phase completes when an active `IRDY#` and either an active `TRDY#` or `STOP#` is clocked. The read transaction completes when `FRAME#` becomes inactive. In reality, there are numerous transaction terminating conditions defined in section 3.3.3 of the PCI specification that can be initiated by either the master or target (for example, *timeout, abort, retry, disconnect*). For our PCI basic read operation, our goal is to demonstrate how to build a transaction through a set of sequence specifications. Hence, we have chosen to simplify our example and ignore these special terminating cases. We leave it to the reader to modify our example by specifying all terminating conditions.

byte enable requirement

Section 3.3.1 (page 47) of the PCI Local Bus Specification states the following requirement associated with a read transaction:

> *The C/BE# output buffers must remain enabled (for both read and writes) for the first clock of the data phase through the end of the transaction.*

Example 3-51 demonstrates how to specify a PCI basic read transaction as specified with the `C/BE#` output buffer requirement. The property PCI_READ_TRANSACTION begins with an address phase (that is, SERE_RD_ADDR_PHASE). We then specify a sequence that describe the initial required turn around cycle (that is, SERE_TURN_AROUND), which occurs the first clock after the address phase. Then, the `C/BE#` signals remain unchanged throughout the remaining data phase `(cbe_n==prev(cbe_n)` throughout SERE_DATA_PHASE.

When specifying protocol requirements, you have the choice of creating a complex property that captures all requirements required for the transaction—or partitioning the different requirements of the transaction into a set of simpler properties. For example, for simplicity we decided not to specify the read transaction latency requirements in Example 3-51 for either the bus target or master (as defined in section 3.5 of the PCI specification). Hence, you could either modify our assertion example by directly writing in the additional bus latency requirements, or you could create a separate simpler property for the latency requirements.

Example 3-51 *PSL* PCI basic read transaction

```
`define data_complete ((!trdy_n || !stop_n) && !irdy_n && !devsel_n)
`define end_of_transaction (data_complete && frame_n)
`define adr_turn_around (trdy_n & !irdy_n)
`define data_tranfer (!trdy_n && !irdy_n && !devsel_n && !frame_n)
`define wait_state ((trdy_n || irdy_n) && !devsel_n)
`define cbe_stable (cbe_n==prev(cbe_n))
`define read_cmd ((cbe_n == `IO_READ) || \
                  (cbe_n == `MEM_READ) || \
                  (cbe_n == `CONFIG_RD) || \
                  (cbe_n == `MEM_RD_MULTIP) || \
                  (cbe_n == `MEM_RD_LINE))

sequence SERE_RD_ADDR_PHASE = {frame_n; !frame_n && read_cmd};
sequence SERE_TURN_AROUND = {adr_turn_around};
sequence SERE_DATA_TRANSFER = {{wait_state[*];data_transfer}[1:inf]}
sequence SERE_END_OF_TRANSFER = {data_complete && frame_n};
sequence SERE_DATA_PHASE =
  {
    {{SERE_DATA_TRANSFER};{SERE_END_OF_TRANSFER}} && {cbe_stable}
  };

property PCI_READ_TRANSACTION =
  always {SERE_RD_ADDR_PHASE} |=> {SERE_TURN_AROUND; SERE_DATA_PHASE}
    abort !rst_n @(posedge clk);

assert PCI_READ_TRANSACTION;
```

Note for this example we are using the PSL *sequence length-matching AND operator* (&&). Hence, this enables us to check throughout the data phase that `C/BE#` is stable throughout the *data transfer* and *end of transfer* sequence.

3.7 Summary

In this chapter, we introduced general concepts related to property specification. We then applied these concepts as we introduced emerging specification standards, which included the Accellera PSL property specification language proposal [Accellera PSL-1.0 2003], the Open Verification Library [Accellera OVL 2003], and SystemVerilog 3.1 assertion constructs [Accellera SystemVerilog-3.1 2003]. Each of the assertion standards we discuss has its own merits. Our objective is to help the engineer understand the advantages (and limitations) of the various assertion forms and their usage model. This will prepare readers to select appropriate specification forms that suit their needs (or preferences). Finally, we demonstrated a process of translating a set of natural language requirements for the Peripheral Component Interconnect (PCI) specification into a set of properties.

CHAPTER

4

PLI-BASED ASSERTIONS

In this chapter, we demonstrate how to create a set of Verilog Programming Language Interface (PLI) assertions. The PLI is a user-programmable, procedural interface that provides a means for interfacing *C* applications with a commercial Verilog simulator. The IEEE 1364-1995 and 1364-2001 standards contain three implementations of PLI library routines. These include the initial OVI *PLI 1.0* standard, which consists of the first generation **TF** and second generation **ACC** libraries, and the later OVI *PLI 2.0* standard, which consists of the third generation of PLI routines, the **VPI** library. The VPI library is a super set of the TF and ACC routines that provides additional capability and simplified syntax and semantics. At the time of this writing, not all commercial simulators support the PLI 2.0 VPI standard. Therefore, all examples in this chapter are coded using the older *PLI 1.0* standard to ensure compatibility with every reader's simulator. We encourage you to implement your PLI-based assertion methodology using the newer *PLI 2.0* standard if your simulator supports the VPI routines. *The Verilog® PLI Handbook* by Stuart Sutherland [2002] is a comprehensive reference manual and guide for learning both the *PLI 1.0* and *2.0* standards.

Even within the PLI 1.0, interpretations of the PLI standard differ between vendors, and vary from what is described in books that attempt to explain the PLI. PLI-based assertion users report that most of their porting problems when testing another vendor's simulator are not in the Verilog text itself, but in their assertion PLI interface. One example is that the status of the parameters seen by the *checktf* varies between Verilog simulation vendors (*checktf* routines are discussed later in this chapter).

Another consideration when implementing a PLI-based assertion solution is the PLI's impact on simulation. A negative impact on simulation can occur if the PLI-based assertion is called at every visit through procedural code to perform its check, particularly when the assertion is not violated.

Broad use of PLI-based assertions places a participating project in the middle of complex portability issues. Hence, we recommend that projects limit their use of PLI-based assertions to a very small set.

PLI-based assertion library

This chapter introduces concepts that enable designers to implement their own PLI-based assertion library. PLI-based assertions; unlike the concurrent assertion constructs introduced in Chapter 3, "Specifying RTL Properties"; may be used directly in procedural code. However, teams must address two issues to prevent false firing of procedural assertions during event-driven simulation. The first issue concerns the potential for evaluating the same procedural blocks multiple times within a *single* simulation time slot. In other words, the transient behavior of variables within a given time slot, prior to reaching a steady-state, could trigger a procedural assertion. The second issue concerns the transient behavior of variables across *multiple* time slots. In other words, generally the engineer is only interested in checking a procedural assertion at a clock edge (that is, the cycle-based semantics of assertions). This is problematic for procedural blocks that are not triggered by a clocking event (for example, modeling combinational logic between sequential elements). In this chapter, we demonstrate the false firing problem encountered by PLI-based procedural assertions in event-driven simulation, and then we present techniques to prevent these errors.

4.1 Procedural assertions

In Chapter 3, "Specifying RTL Properties", we introduced the OVL. This library consists of a set of assertion modules that concurrently validate an RTL expression at every edge of a sample clock. While *OVL concurrent assertions* prevent false firings by sampling the assertion test expression at a clock edge, *procedural assertions* are only checked during procedural visits through the code. In this section, we present some of the strengths of procedural assertions; for example, expressiveness and convenience. We also discuss the weaknesses of procedural assertions; for example, over constraining and false firing if not properly constructed.

over-constraining procedural assertions

Experience has demonstrated that, when compared to declarative forms of assertions, procedural assertions that are deeply nested within RTL case and if statements run a higher risk of being over constrained. When this is the case, they can miss a bug. That is not to say declarative assertions are immune to over constraint. However, when designers embed assertions deeply within RTL code, they must seriously consider the effect on the assertion for all conditional expressions related to the nested case and if statements. Particularly as the design undergoes changes.

Example 4-1 demonstrates this point. In this simplified example, the engineer is using a PLI task to validate that the three one-bit variables (a, b, and c) are mutually exclusive. However, the mutually exclusive property only validates when d is true. If the true property of these variables is that they should *always* be mutually exclusive, then there is a potential to miss an error in the design during verification if d is not true during the previous visit. In actual RTL code, unlike this simple code in Example 4-1, assertions deeply nested within case and if statements are often quite complex. Therefore, the designer must be especially alert to the potential for over constraining assertions embedded in procedural code.

Example 4-1 Over constrained procedural assertion

```
always @(a or b or c or d) begin
  :
  if (d)
    $assert_one_hot ({a,b,c});
end
```

Procedural assertion convenience

In spite of the potential problem of over constraining a procedural assertion, designers generally prefer the convenience of expressing assertions procedurally. For example, the assertion expression, created for an OVL concurrent check, can become quite complicated if it is necessary to qualify the assertion with an expression that represents the sensitized path down through the deeply nested procedural code. However, if the designer places the assertion directly in the procedural code, this reduces the amount of required coding.

4.1.1 A simple PLI assertion

Example 4-2 demonstrates the source code for a simple PLI $assert_always() check. This PLI assertion validates the designer's Verilog Boolean expression, which is passed in as its argument, is always true. The PLI *C* code for this assertion is divided into two functions: a *checktf* routine and a *calltf* routine.

Checktf routine

The assert_always_checktf() function, shown in Example 4-2, is automatically called by the simulator before the simulator starts running (that is, prior to simulation time 0). This is either at the Verilog source code compilation time or load time, depending on the simulator. The purpose of the assert_always_checktf() function is to verify that the arguments used in the PLI *system task* are used correctly when instantiated within the designer's RTL (for example, correct number of arguments, or correct expression width).

Example 4-2 PLI checktf routine for $assert_always

```
/*******************************************
* checktf routine to validate arguments
*******************************************/
int assert_always_checktf(char *user_data)
{
  if (tf_nump() != 1)
    tf_error("$assert_always only 1 argument.");
  else if (tf_sizep(1) != 1)
    tf_error("$assert_always argument size!=1");
  return(0);
}
```

The details for the individual PLI routines used in the assert_always_checktf() routine shown in Example 4-12 are as follows:

tf_error

The tf_error() routine, shown in Example 4-2, is similar to the *C* printf() function and can be used to print an error message to the simulator's output window. The tf_error() routine, when called from a user's *checktf* routine, prints an error message and then aborts the simulator process.

tf_nump

The tf_nump() routine, shown in Example 4-2, returns the number of arguments passed into the Verilog-instantiated $assert_always PLI task. If the number of arguments is not equal to one in our example, then the *checktf* routine reports an error prior to the beginning of simulation.

tf_sizep

The tf_sizep() routine, shown in Example 4-2, returns the bit width for an indexed, referenced argument in the instantiated PLI call. The *checktf* routine, illustrated in Example 4-2, checks that the first argument used in the instantiated PLI call, either a variable or an expression, is of size 1 (for example, `$assert_always(a==0)`, in which the first argument is the expression `a==0` and is of size 1). If the bit width of the first argument is not equal to one, then the *checktf* routine reports an error prior to the beginning of simulation.

Example 4-3 PLI calltf routine for $assert_always

```
/*******************************************
* calltf routine to check assertions
*******************************************/
int assert_always_calltf(char *user_data)
{
 /* read current value */
  if (tf_getp(1) == 0) {
    io_printf ("ASSERT ALWAYS ERROR at %s:%s\n",
            tf_strgettime(), tf_spname());
    tf_dofinish(); /* stop simulation */
  }
  return(0);
}
```

Calltf routine

The assert_always_calltf() *calltf* routine, shown in Example 4-3, is invoked each time an $assert_always() PLI task is encountered during the simulator's procedural visit through the Verilog code. This routine validates the user's assertion test expression.

The details for the individual PLI routines used in the assert_always_calltf routine shown in Example 4-2 are as follows:

tf_getp — The tf_getp() routine returns the specific instance's current value for a referenced argument. In our example, the tf_getp(1) is referencing the first argument in the instantiated PLI assertion, which is the Verilog expression we are asserting to be true.

io_printf — The io_printf() routine is similar to the *C* printf() function and can be used to print formatted text for up to twelve arguments. The text message will be printed to both the simulator's output window and the simulator's output log.

tf_strgettime — The tf_strgettime() routine returns the current simulation time as a string.

tf_spname — The tf_spname() routine returns a point to a string that contains the hierarchical path name for the instantiated PLI assertion.

tf_dofinish — The tf_dofinish() routine performs the same function as the Verilog $finish() built-in system task—which is to close all open files and cause the simulator to exit.

Sutherland [2002] provides an excellent description of techniques for linking the user's PLI application to various Verilog simulators.

4.1.2 Assertions within a simulation time slot

procedural assertions may encounter false firings in simulation

Unlike the OVL assertion modules, which sample the Verilog assertion expression at a clock edge, PLI-based procedural assertions validate the assertion expression each time an assertion is encountered during a procedural visit through the code. Hence, procedural assertions run the risk of false firing due to the transient behavior of variable assignments during event-driven simulation.

Example 4-4 illustrates this potential for false firing using a $display system task to report the assertion violation. That is to say, the ordering (or scheduling) of events within the same simulation time slot can cause the procedural always block to execute multiple times. Therefore, the transient behavior of the a and b variables within the same time slot can cause a procedural assertion to fire.

Example 4-4 False firing of procedural assertion

```
always @(a or b) begin
  $display (a^b);
end
```

Figure 4-1 illustrates multiple time slots of the simulator, and then shows the details of the simulator's event queue for time slot 3. For demonstration purposes, the simulation time slots are represented symmetrically, although in practice this would generally not be the case.

Notice the details for the event queue in time slot 3. As simulation progresses, new variable assignments are placed on the event queue and are scheduled for evaluation within the current time slot. This is due to the occurrence of new blocking assignments generated during the current simulation time slot. It is possible that the PLI assertion would fire when a==0 and b==0 during an early evaluation of the procedural code for time slot 3. However, later within this same time slot, variable a is schedule for evaluation with a new value of 1. Any previous assertion violations are no longer valid for this time slot. Hence, to prevent false firings, the PLI routine must be constructed in such a way that it will wait to evaluate the assertion expression after the transient behavior of the simulated variables settles out, which is at the end of simulation time slot 3.

Figure 4-1 **Simulation event queue**

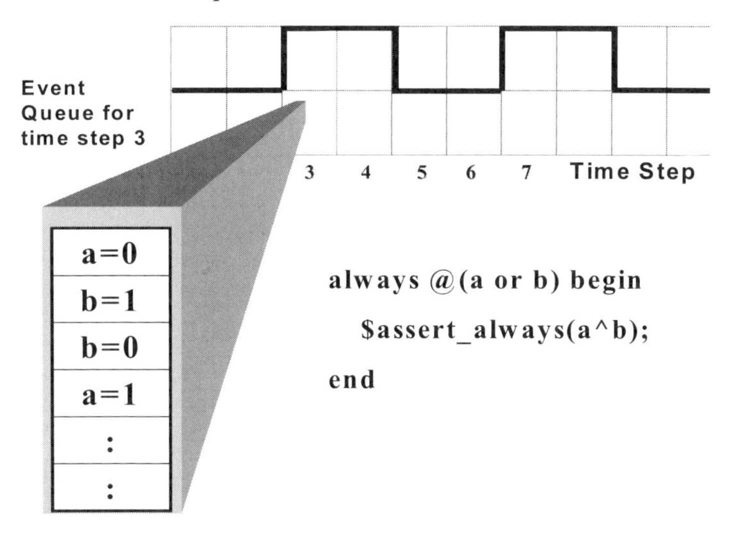

Example 4-5 illustrates an enhancement to the $assert_always() PLI code previously shown in Example 4-2. This enhancement prevents a false firing of the assertion by rescheduling the assertion evaluation to occur at the end of the time slot (after all transient evaluations have settled out).

The PLI *mistf* routine is used to execute a *C* application after a miscellaneous event occurs during simulation. Miscellaneous simulation events include: *end of compilation*, *end of simulation*, *end of a simulation time slot*, as well as many other simulation events. (See Sutherland [2002] for additional details on PLI *misctf* routines and simulation events.) In the following section, we will demonstrate how to perform a callback at the end of the current simulation time slot using a tf_rosynchronize routine combined with a *misctf* routine.

tf_rosynchronize

The tf_rosynchronize() routine shown in the assert_always_calltf() code in Example 4-5 schedules a callback to a user-defined *mistf* routine (for example, assert_always_mistf()). This callback occurs at the end of the current simulation time slot.

The simulator generates the `reason` constant, shown in Example 4-5, and passes it to the *mistf* routine during its simulation call. If the *mistf* routine is called due to an *end of a time slot event*, the `reason` integer is set to a constant value of `REASON_ROSYNCH`, as defined in the simulator's *verisuser.h* file. If the assert_always_misctf routine is called for any simulation event, other than a `REASON_ROSYNCH`, then the *mistf* routine exits without performing the assertion check. By checking the assertion only during a REASON_ROSYNCH event, false firing of assertions within a simulation time slot is eliminated.:

Example 4-5 $assert_always with callback at end of current time slot

```
/*******************************************
* checktf routine to validate arguments
*******************************************/
int assert_always_checktf(char *user_data)
{
  if (tf_nump() != 1)
    tf_error("$assert_always only 1 argument.");
  else if (tf_sizep(1) != 1)
    tf_error("$assert_always argument size!=1");
  return(0);
}
/********************************************
* calltf routine to schedule callback
********************************************/
int assert_always_calltf(char *user_data)
{
  tf_rosynchronize();
  return(0);
}/*******************************************
 * misctf routine to check assertion
 *******************************************/
int assert_always_misctf(char *user_data,
                         int reason, int paramvc)
{
  if (reason != REASON_ROSYNCH)
    return(0);

/* read current value */
  if (tf_getp(1) == 0) {
    io_printf ("ASSERT ALWAYS ERROR at %s:%s\n",
            tf_strgettime(), tf_spname());
    tf_dofinish(); /* stop simulation */
  }
  return (0);
}
```

Nested PLI assertion problem

In the previous section, we demonstrated how to construct a PLI routine that can safely prevent a false firing within the procedural code by scheduling a callback evaluation that is performed at the end of the simulation time slot in which the PLI routine was called. However, this assertion can still encounter problems. Refer to and note that if there is a transient behavior of the c variable that initially causes the procedural block to execute by assigning one to c, this would schedule a callback at the end of the current simulation time slot. However, if the variables within the simulation reached a steady state and the final value of the variable c is zero, then the PLI assertion really should not have been visited during the current time slot. This could cause a false firing of the assertion.

Example 4-6 False firing of procedural assertion

```
always @(a or b or c) begin
  if (c)
    $assert_always (a^b);
end
```

To prevent a false firing, we recommend that you associate a sampling clock with the PLI-based assertions that will cause the assertion to be scheduled for a future evaluation on an edge event associated with the change of the clock. This technique is discussed in the next sections.

4.1.3 Assertions across simulation time slots

Often, procedural blocks represent purely combinational logic that is bounded between sequential elements that are described in separate clocked procedural blocks. However, the designer generally focuses on the cycle-based semantics for any assertions within these combinational blocks. In this section, we demonstrate a technique of associating a clock with a PLI-based assertions to achieve assertion cycle-based semantics.

false firing example across multiple time slots

Figure 4-2 illustrates seven time slots of the simulator for the procedural block shown in Example 4-6. For demonstration purposes, the simulation time slots are represented symmetrically, although in practice this would generally not be the case. At time slots 2 and 3, no assertion fires since the a and b variables are mutually exclusive. For time slots 1, 4, 5, and 7; the procedural assertion fires due to the non-mutual exclusivity of the a and b variables. At time slot 6, the c variable should prevent evaluation of the procedural assertion. However, if the combinational logic feeds into sequential elements, we are generally only interested in the cycle-based semantics for these procedural assertions. In other words, we should evaluate the procedural assertion at the rising edge of the clock.

Figure 4-2 Assertions across multiple simulation time slots

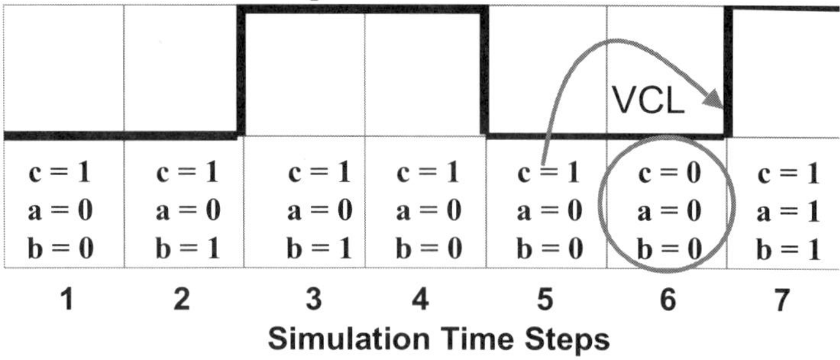

cycle-based semantics for procedural assertions

Example 4-7 illustrates a VCL version of the PLI $assert_always_ck assertion that attempts to prevent a false firing of the assertion across multiple simulation time slots by introducing a sampling clock to the assertion. Note that the clock is not part of the procedural block's event sensitivity list (for example, `@(a or b or c))`. In other words, we do not want to change the procedural block's simulation behavior by having it trigger on the rising and falling edge of the clock. However, we want to ensure that the assertion schedules an evaluation of the PLI routine on the next rising edge of a clock

Example 4-7 Procedural assertion with a Value Change Link on the clock

```
always @(a or b or c) begin
  :
  if (c)
    $assert_always_ck (ck, a^b);
end
```

a simple clock edge detection approach will not work

Note that we cannot be assured that the procedural code in Example 4-7 will execute prior to and after a rising edge of the clock. For example, the variables in the event sensitivity list do not change values between time slot 2 and time slot 3 (as shown in Figure 4-2), which would mean that the procedural block would not be evaluated during time slot 3. Hence, we cannot simply detect an edge of the clock for assertion evaluation by keeping a record of the clock value within the PLI routine for previous procedural visits of the assertion. However, by using PLI Value Change Link (VCL) routines, the simulator can schedule an evaluation of the assertion to occur on the appropriate edge of the clock.

Controlling assertion evaluations by a clock

using the VCL to monitor clock edges for assertions

The PLI assertion shown in Example 4-7 can be constructed utilizing the VCL routines within the ACC library. These VCL routines allow the Verilog simulator to schedule an evaluation of the user's PLI *consumer* routine by monitoring a specific simulation object for value changes. For our example, this enables us to delay the evaluation of the assertion until a positive edge of the clock occurs at the start of steps 3 and 7 as shown in Figure 4-2.

Example 4-8 (below) demonstrates $assert_always_ck assertion coding that uses the VCL to provide cycle-based semantics for the assertion evaluation.

Each time the $assert_always_ck task is encountered during a procedural visit through the Verilog code, the *calltf* routine, shown in Example 4-8, activates a VCL monitor for the assertion clock. The `assertion->active` flag tracks previously activated VCL monitors to prevent activating multiple monitors between clock edges.

The VCL *consumer* routine is called whenever the assertion's clock changes values (that is, the *clock object*). This routine retrieves the previously recorded clock value, captured in the *calltf* routine, and uses this value to determine rising edges of the clock. If a rising edge is detected, the *consumer* routine retrieves and validates the assertion test expression. After validating the assertion, the *consumer* routine removes the VCL monitor from the clock object. Future evaluations of the assertion now depend on encountering new procedural visits to the PLI assertion task; at which time, a new VCL monitor is activated for the clock object.

The details for the individual PLI routines used in Example 4-8 are as follows:

acc_initialize

The acc_initialize() routine is required when using the ACC library routines. This function resets the ACC environment to default values.

tf_setworkarea

The tf_setworkarea() routine enables a specific instance of a system task to store a pointer for working area PLI application data. The stored data can be retrieved for use within a *mistf* or *consumer* routine. In our $assert_always_ck example, we must store one handle for each instance to the clock object, which will be used during the VCL evaluation of the assertion's *consumer* routine. In addition, we must store a string that represents the instance path and the value of the clock during the last PLI call, to determine a rising edge of the clock within the *consumer* routine.

Example 4-8 $assert_always_ck with VCL callback to consumer routine

```
typedef struct always_t {
  int active;
  handle clk_handle;
  char *instance;
  int clk;
} Always;
/*******************************************
* checktf routine to validate arguments
*******************************************/
int assert_always_ck_checktf(char *user_data)
{
  if (tf_nump() != 1)
    tf_error("$assert_always_ck only 1 argument.");
  else if (tf_sizep(1) != 1)
    tf_error("$assert_always_ck argument size!=1");
  return(0);
}
/*********************************************
* calltf routine setup VCL for clock
*********************************************/
int assert_always_ck_calltf(char *user_data)
{
    struct always_t * assertion;

    acc_initialize();
    assertion = (Always *) tf_getworkarea();

    if (assertion == NULL) {
      assertion = (Always *) calloc(1,sizeof(Always));
      assertion->clk_handle = acc_handle_tfarg(1);
      assertion->instance = tf_getinstance();
      assertion->active = 0;
    }
    assertion->clk = tf_getp(1) & 1;
    tf_setworkarea((char *) assertion);
    if (assertion->active == 0) {
      assertion->active = 1;
     acc_vcl_add(assertion->clk_handle, assert_always_ck_consumer,
                 char *) vcl_verilog_logic);
    }
}
/*******************************************
* checktf routine to validate arguments
*******************************************/
int assert_always_ck_checktf(char *user_data)
{
  if (tf_nump() != 1)
    tf_error("$assert_always_ck only 1 argument.");
  else if (tf_sizep(1) != 1)
    tf_error("$assert_always_ck argument size!=1");
  return(0);
}
```

Example 4-8 $assert_always_ck with VCL callback to consumer routine

```
/********************************************
* calltf routine setup VCL for clock
*********************************************/
int assert_always_ck_calltf(char *user_data)
{
    struct always_t * assertion;

    acc_initialize();
    assertion = (Always *) tf_getworkarea();

    if (assertion == NULL) {
      assertion = (Always *) calloc(1, sizeof(Always));
      assertion->clk_handle = acc_handle_tfarg(1);
      assertion->instance = tf_getinstance();
      assertion->active = 0;
    }
    assertion->clk = tf_getp(1) & 1;
    tf_setworkarea((char *) assertion);
    if (assertion->active == 0) {
      assertion->active = 1;
      acc_vcl_add(assertion->clk_handle,
             assert_always_ck_consumer, (char *) vcl_verilog_logic);
    }
    acc_close();
    return(0);
}
/********************************************
 * consumer routine to perform check
 ********************************************/
int assert_always_ck_consumer(p_vc_record vc_record)
{
  int clk;
  struct always_t *assertion;
  assertion = (Always *) vc_record->user_data;
  switch (vc_record->vc_reason) {
    case logic_value_change:
    case sregister_value_change: {
      clk = tf_igetp(1, assertion->instance) & 1;
      if (assertion->clk==0 && clk==1) {
       if (tf_igetp(2, assertion->instance)==0 {
         io_printf ("ASSERT ERROR at %s:%s\n",
             tf_strgettime(), tf_ispname(assertion->instance));
         tf_dofinish(); /* stop simulation */
       }
       assertion->active = 0;
       acc_vcl_delete(assertion->clk_handle,
                      assert_always_ck_consume,(char *) assertion,
                      vcl_verilog_logic);
    }
    assertion->clk = clk;
  }
 }
 return (0);
}
```

tf_getworkarea The tf_getworkarea() routine retrieves a pointer to previously stored work area data for each PLI instance.

acc_handle_tfarg The acc_handle_tfarg() routine returns a handle for the object referenced by the argument number in the instantiated PLI system task. In our example, we reference the first argument in the instantiated PLI call, which is the clock.

tf_getinstance The tf_getinstance() routine returns a pointer to the instance of a PLI system task. In our example, this instance pointer is used in our *consumer* routine to obtain the values of the assertion test expression and clock, as well as hierarchical path name to the instantiated PLI assertion.

acc_vcl_add The acc_vcl_add() routine adds a VCL monitor for the assertion clock defined by our specific, instantiated PLI assertion. When the clock changes values, the *consumer* routine executes.

acc_close The acc_close() routine frees all memory allocated by the acc_initialize() routine and resets all configuration parameters to their default values.

acc_vcl_delete The acc_vcl_delete() routine removes any previously activated VCL monitors on a specific instance object. In our example, we want to stop monitoring the assertion clock as soon as a rising edge of the clock occurs.

tf_igetp The tf_igetp() routine is similar to the previously-defined tf_getp routine. This routine requires a pointer to the instance of the PLI assertion for which we wish to retrieve the argument value. The instance pointer was captured by the *calltf* routine and stored within the instance's working area.

tf_ispname The tf_ispname() routine returns a pointer to a string that contains the hierarchical path name for the instantiated PLI assertion. This routine requires a pointer to the instance of the PLI assertion.The tf_igetp() routine is similar to the previously defined tf_getp routine. This routine requires a pointer to the instance of the PLI assertion for which we wish to retrieve the argument value. The instance pointer was captured by the *calltf* routine and stored within the instance's working area.

4.1.4 False firing across multiple time slots

The problem with the VCL PLI assertion approach, shown in Example 4-7, is that the procedural path down to the assertion is no longer valid (or sensitized) at simulation time slot 6, as demonstrated in Figure 4-2, since the c variable is false at this point in time. Therefore, the $assert_always_ck should not evaluate at the next clock edge. However, a VCL was previously activated due to a visit by the assertion at time slots 3, 4 and 5 to monitor the clock. To solve this problem, the procedural always block shown in Example 4-7 requires a method of deleting any previously scheduled VCL clock monitors for all PLI assertion routines during each new visit to the procedural block.

Example 4-9 demonstrates one technique that could be used to delete all previously scheduled VCL clock monitors within the labeled procedural block. When an $assert_always_ck assertion is encountered during a procedural visit, the $assert_always_ck PLI code records the activation of an assertion in a globally accessible data structure, which identifies the specific assertion in the labeled procedural block that was activated. The $assert_delete user-defined task has access to this data structure. Upon future visits to the procedural block, any previously activated VCL PLI assertion is disabled by the $assert_delete task, which calls an acc_vcl_delete() routine for any previously initiated VCL assertions.

Example 4-9 Procedural assertion with a Value Change Link on the clock

```
always @(a or b or c) begin : my_block
  $assert_delete();
  :
  if (c)
    $assert_always_ck (ck, a^b);
end
```

In general, procedural assertions can be deeply nested within case and if statements and should only be evaluated when the procedural path down to the assertion is valid during the last time slot prior to the clock. Otherwise, a false firing of the assertion can occur. Semantically, the PLI procedural assertion should behave as shown in Example 4-10, below.

Example 4-10 Required semantics for safe RTL assertion

```
reg test_assertion;
always @(a or b or c) begin
  test_assertion = 1'b1
   :
  if (c) test_assertion = a^b;
end
assert_always ovl_assert (ck, reset_n,
                          test_assertion);
```

For this example, we introduce a variable to test the assertion within the procedural code. The variable is initialized to true at the beginning of the procedural code. Later, the test expression is set based on the evaluation of the a^b expression. To complete our example, an OVL assert_always module is instantiated to sample the *test_assertion* variable on a rising edge of a clock.

Hence, this example illustrates the semantics required when creating a set of PLI-based assertions.

Note that the OVL assertion module has a reset signal as an argument. The PLI-based $assert_always_ck task in our Example 4-8 and Example 4-9 could be modified to include a reset signal as an argument. In addition, an optional assertion message could be passed in as an argument, which can be displayed if the assertion fires. The PLI application determines if the optional parameter has been passed in by calling the tf_nump() routine. (Refer to "tf_nump" on page 4-104.) If the number of arguments in our example is equal to three, then only the *clock*, *reset*, and assertion *test expression* have been passed in as arguments. If the number of arguments is four, then the optional error message was specified in the instantiated PLI task. The tf_typep() routine can be used to ensure that the optional argument is a literal string (for example, `tf_type(4)==TF_STRING`), which is used in the *checktf* routine for the $assert_always_ck task. The assertion would be instantiated as shown in Example 4-11, below.

Example 4-11 PLI assert always check

```
$assert_always_ck (ck, reset_n, expression,
            "My optional assertion error message" )
```

4.2 PLI-based assertion library

In the previous section, we demonstrated a PLI-based implementation for an $assert_always procedural assertion. You can construct a library of PLI procedural assertions, with functionality similar to the OVL described in Chapter 3, "Specifying RTL Properties", just as we constructed the previous $assert_always assertion example.

We suggest you begin by constructing a simple procedural $assert_error task. This is useful for the default alternative error branch in case statements, or the else error branch in if statements. Example 4-12 (below) demonstrates the use of a simple $assert_error task within a case statement.

Example 4-12 $assert_never within a CASE default branch

```
case({a,b}) begin
  2'b01: s = c+4'b0001;
  2'b10: s = c-4'b0001;
  default: $assert_error("Case default error");
endcase
```

When creating a PLI-based assertion library, the target type of procedural block must be considered (for example, a clocked procedural block versus a non-clocked procedural block). Hence, the designer might decided to implement two sets of PLI-based assertions, as shown in Example 4-13. Note that the first assertion applies to clocked procedural block, while the second applies to a non-clocked procedural block.

Example 4-13 Clocked versus non-clocked procedural PLI assertions

```
always @(posedge ck) begin
  $assert_always (a^b);
end

always @(a or b) begin
  $assert_delete();
  $assert_always_ck (ck, a^b);
end
```

In the following section, we demonstrate another example of implementing a PLI-based assertion that could be included in the designer's PLI-based assertion library.

4.2.1 Assert quiescent state

After you have tried your hand at the PLI-based assert_always, the next one you will want to attempt is the assert quiescent state, which is useful for validating state machines after the completion of a transaction or sequence of events. This PLI-routine performs a new and powerful function not supported by the released version of the OVL assert_quiescent_state module (that is, our PLI routine executes an automatic callback at the *end of simulation* to perform a consistency checking). The *end of simulation* assert quiescent state evaluation is useful for identifying bus transactions that have stalled, deadlock states, and inconsistent behavior between multiple state machines.

Example 4-14 PLI quiescent state assertion

```
#ifndef MAXINT
#define MAXINT 32
#endif
#ifndef NUMWORDS
#define NUMWORDS(x) (x-1)/MAXINT
#endif
/********************************************
 * checktf application to validate interface
 ********************************************/
int quiescent_state_checktf(char *user_data)
{
  s_tfexprinfo expr_info;
  int current_size, end_size;
  if (tf_nump() != 2)
    tf_error(
      "$assert_quiescent_state requires 2 arguments");
  current_size = tf_sizep(1);
  end_size = tf_sizep(2);
  if (current_size != end_size)
    tf_error(
      "$assert_quiescent_state arg(1) and arg(2) size mismatch.");
    return(0);
}

/****************************************
 * calltf routine
 ****************************************/
int quiescent_state_calltf(char *user_data)
{
  tf_rosynchronize();
  return(0);
}
/************************************************
 * misctf routine to perform end of sim. check
 ************************************************/
int quiescent_state_misctf(char *user_data,
                        int reason, int paramvc)
{
  int current_quiescent_state,
  int end_quiescent_state;
  int current_high, end_high;
  int i, size, ok=1;
  s_tfexprinfo current_expr_info, end_expr_info;

  if (reason != REASON_FINISH)
    return(0);
  size = tf_sizep(1);
  if (size <= 32) {
    current_quiescent_state = tf_getp(1);
    end_quiescent_state = tf_getp(2);
    if (current_quiescent_state !=
                            end_quiescent_state)
      ok = 0;
  }
```

Example 4-14 PLI quiescent state assertion

```
    else if (size <= 64) {
      current_quiescent_state =
                    tf_getlongp(&current_high, 1);
      end_quiescent_state =
                    tf_getlongp(&end_high, 2);     if
  (!((current_quiescent_state ==
            end_quiescent_state) &&
          (current_high == end_high)))
        ok = 0;
    }
    else {
      (void) tf_exprinfo(1, &current_expr_info);
      (void) tf_exprinfo(2, &end_expr_info);
      for (i=
        NUMWORDS(current_expr_info.expr_vec_size);
        i >= 0; i-=1) {
        if (current_expr_info.expr_value_p[i].avalbits
            != end_expr_info.expr_value_p[i].avalbits) {
          ok = 0;
          break;
        }
      }
    }
    if (ok==0) {
      io_printf ("$assert_quiescent_state error: %s : %s\n",
                          tf_spname(), tf_getp(1));
    }
    return (0);
  }
```

The details for the individual PLI routines used in Example 4-14 are as follows:

end of simulation automatic check

The `reason` constant is generated by the simulator and passed to the *mistf* routine during its simulation call. If the *mistf* routine is called due to an *end of a simulation event*, the `reason` integer is set to a constant value of `REASON_FINISH`, as defined in the simulator's *verisuser.h* file.

processing arguments greater than 32 bits

If the quiescent state arguments are greater than 32 bits, but less than 64 bits, then you must use the tf_getlongp() routine to retrieve the argument's value. If the quiescent state arguments are greater than 32 bits, then you must use the tf_exprinfo() function to obtain detailed information about the system task's argument. This information is retrieved into an s_tf_exprinfo structure. The value of the argument is stored in an array of s_vecval structures and is referenced by the expr_value_p pointer from the s_tf_exprinfo structure. The quiescent state is validated by looping through the s_vecval structures and comparing the existing expression value (argument 1) with the expected end of simulation value (argument 2).

Example 4-15 OVL assert_quiescent_state enhanced with PLI task

```
module assert_quiescent_state (clk, reset_n,
     state_expr, check_value, sample_event);
// rtl_synthesis template
  parameter severity_level = 0;
  parameter width=1;
  parameter options = 0;
  parameter msg="VIOLATION";
  input clk, reset_n, sample_event;
  input [width-1:0] state_expr, check_value;

//rtl_synthesis translate_off
`ifdef ASSERT_ON
  parameter assert_name = "ASSERT_QUIESCENT_STATE";

  integer error_count;
  initial error_count = 0;

  `include "ovl_task.h"
  `ifdef ASSERT_INIT_MSG
    initial ovl_init_msg;
  `endif

  reg r_sample_event;
  initial r_sample_event=1'b0;

  always @(posedge clk)
    r_sample_event <= sample_event;

  reg r_PLI_active;
  initial r_PLI_active=1'b0;
  always @(posedge clk)
    if (r_PLI_active==1'b0) begin
      r_PLI_active=1'b1;
      $assert_quiescent_state (state_expr, check_value);
    end
  always @(posedge clk) begin
    `ifdef ASSERT_GLOBAL_RESET
      if (`ASSERT_GLOBAL_RESET != 1'b0)
    `else
      if (reset_n != 0)
    `endif
      begin
        if ((r_sample_event == 1'b0 && sample_event == 1'b1) &&
             (state_expr  != check_value))
        begin
          ovl_error("");
        end
      end
  end
`endif
//rtl_synthesis translate_on
endmodule
```

OVL enhancement with PLI call In addition to building your own PLI-based assertion library, you can use PLI assertions to enhance the existing OVL modules. Example 4-15 (below) shows our enhancement to the OVL assert_quiescent_state module, which adds the PLI procedural $assert_quiescent_state assertion to the OVL module. This PLI routine validates the user's specified state automatically at the end of simulation. The r_PLI_active signal enables us to visit the PLI routine once, setting up the callback to the *misctf* routine to occur at an *end of simulation event*, and then ignore the PLI call for the remainder of the simulation run.

The designer might decide to instantiate an enhanced version of the OVL assertion shown below in Example 4-16 to check for only an *end of simulation* violation with our new PLI routine. In this case, the OVL `sample_event` argument is set to 1'b0. When simulation completes, the PLI's *mistf* routine is invoked to automatically check the `controller_state` bits 7 through 0 for the value specified by the `'CNTRL_START_STATE` macro. If any state other than the expected state is encountered at the end of simulation, the assertion will fire.

Example 4-16 OVL instantiated assertion to check for an end of simulation violation

```
assert_quiescent_state valid_state (ck,
      reset_n, controller_state[7:0],
      'CNTRL_START_STATE, 1'b0);
```

4.3 Summary

In this chapter, we demonstrated how to create a set of Verilog Programming Language Interface (PLI) assertions. We then discussed considerations to take when implementing a PLI-based assertion solution and their impact on simulation. For example, a negative impact on simulation can occur if the PLI-based assertion is called at every visit through procedural code to perform its check, particularly when the assertion is not violated. We then presented some of the strengths of PLI-based procedural assertions; for example, expressiveness and convenience. We also discussed the weaknesses of procedural assertions; for example, over constraining and false firing if not properly constructed. Broad use of PLI-based assertions places a project in the middle of complex portability issues. Hence, we recommend you limit the use of PLI-based assertions to a very small set.

CHAPTER 5

FUNCTIONAL COVERAGE

As the complexity of today's ASIC designs continues to increase, the challenge of verifying these designs intensifies at an even greater rate. Advances in this discipline have resulted in many sophisticated tools and approaches that aid engineers in verifying complex ASIC designs. However, the age-old question of *when is the verification job done*, remains one of the most difficult questions to answer. Consider random test generators, which are heavily used to generate simulation stimulus on-the-fly. At issue is knowing which portions of a design are repeatedly exercised and which portions are not touched at all. Or more fundamentally, exactly *what* functionality has been exercised using these techniques. Historically, answering these questions has been problematic. This has led to the development of various coverage metrics ranging from code coverage (used to identify unexercised lines of code) to functional coverage (used to identify key functionality that has not been explored).

When specifying functional coverage, there is actually a spectrum ranging from higher-level architectural functional coverage models down to lower-level functional coverage models for RTL implementation. For example, specifying that a certain sequence related to a bus transaction must be encountered during the course of verification would be a higher form of functional coverage— and we provide examples of transaction forms of specification that can be used for functional coverage in Section 5.7 "AHB example" on page 156. Conversely, specifying that a specific FIFO counter must reach its maximum value at some point during simulation is an example of lower-level functional coverage for RTL implementation. There are many excellent commercial products available focusing on *higher levels of functional coverage*. However, the verification team building higher-level functional coverage models often neglects many interesting user-defined *functional coverage points for RTL implementation*. In fact, verification engineers generally lack the depth of knowledge of the implementation that would allow them to specify important corner case conditions (like FIFO pointer overflow or underflow). Hence, a good functional coverage methodology combines higher-level forms of functional coverage related to the specification with lower-level functional coverage for corner cases in the RTL implementation.

This chapter explores verification approaches that provide feedback about what *has* been tested and what *has not* been tested. We begin by building a knowledge base for understanding testing approaches, and from there, we lead into coverage techniques and build a case for user-defined functional coverage. After laying this foundation, we concentrate on our central topic, which is adopting an effective functional coverage methodology that provides pertinent feedback for RTL implementation. Finally, we explore available RTL implementation functional coverage technologies and close this chapter with actual functional coverage examples.

The reader might wonder, *why are you talking about functional coverage in a book focused on assertions*? In fact, the two topics are related. For example, a property language, such as PSL, can specify assertions (which monitor and report *undesirable* behavior) as well as functional coverage (which monitors and reports *desirable* behavior that must occur for the verification process to be complete). Please note that our emphasis in this book is on RTL *implementation-level* functional coverage, as opposed to system or *specification-level* functional coverage. There are numerous commercial solutions available that address specification-level functional coverage as well a literature written on this topic [Bergeron 2003]. And while all levels of functional coverage are valuable, the significance of implementation-level functional coverage specified by the designer is often overlooked. Hence, we have decided to expend a little more effort discussing techniques and methodologies around implementation-level functional coverage.

5.1 Verification approaches

black-box verification

The typical approach in design verification is through *black-box verification*. As discussed in Chapter 1, "Introduction", with this approach, a team creates a model of a design written in a hardware description language that is instantiated in a system-level model (testbench) that drives stimulus to the design under verification (DUV) and provides a mechanism for observing and validating the output responses. Recall that this approach has limited observability and controllability. And these aspects become worse as the size of the design grows,

white-box verification

White-box testing can effectively complement traditional black-box testing by adding intimate knowledge of the internal implementation of the design. Recall from our discussion in Chapter 1 that instead of simply stimulating the external ports of a black-box using knowledge of the expected external stimulus and response, white-box testing adds a dimension of understanding that looks into how the external stimulus is used and how the external response is generated. With this insight, an engineer can also ensure that internal features operate correctly.

As discussed earlier in this book, we add assertions to monitor internal behavior. For example, an assertion can monitor a state machine to ensure that it is always one-hot. In the presence of undesirable behavior, an assertion identifies the error at its source. Additionally, assertions can validate designer assumptions and areas of concern.

Another white-box testing method adds internal or implementation level *functional coverage* to the RTL. This also extends the verification environment to more areas of the design and does not confine observation to external ports of the DUV. *Functional coverage* monitors desirable (and expected) behavior in the same way that *assertions* monitor undesirable (and unexpected) behavior. Thus white-box testing; with assertions, functional coverage, or a combination of both; provides an effective verification method, which is well documented by Kantrowitz and Noack [1996], Taylor et al. [1998], Bentley [2001], Bergeron [2003], Lacish et al. [2002], FoCs [2003], Ziv [2002a], Ziv [2002b], Betts et al. [2002] and Bening and Foster [2001]. However, using only white-box testing misses system-level failures that the testbench, which is used in black-box testing, can detect. Consequently, an ideal methodology combines white-box and black-box testing approaches.

gray-box verification

The term *gray-box testing* is often used when we blend black-box verification methods with elements of white-box verification. By using this approach, the verification coverage can be greatly increased.

5.2 Understanding coverage

Functional verification, in general, is a process that demonstrates that the RTL implementation satisfies all requirements established in the *specification* and *design/architect* phases of the design process (see Section 1.4 "Phases of the design process" on page

14), while not exhibiting any unexpected behavior. Ultimately, the only thing that matters in functional verification is high coverage; that is, ideally we would like to explore all combinations of input values with respect to all possible sequences of internal state. Without high coverage, corner cases go unexplored, which can result in functional bugs within the silicon. Traditionally, vector-based verification techniques (such as simulation, acceleration, and emulation) have been the primary processes used for design validation, coupled with coverage techniques to expose unverified portions of the design.

To ensure that a design is correct when using traditional simulation techniques, the design must be exercised with all possible sequences of input and register states. While this is possible on smaller designs, today's large, complex designs make this method impractical. Additionally, formal techniques such as model checking can be used to exhaustively verify correct functional behavior; however, this technique does not scale to large designs. In view of the coverage limitations of available techniques, we must devise other methods that enable us to gather coverage data and answer the question: *When is the verification job done?* To address the question adequately, we must know what we *have not* verified (holes) and what functionality we *have* verified (related to the specification).

5.2.1 Controllability versus observability

Fundamental to the discussion of coverage is understanding the concepts of controllability and a observability. *Controllability* refers to the ability to stimulate a specific line of code or structure within the design. Note that, while in theory a testbench has high controllability of the input bus of its device under verification, it can have low controllability of an internal point. *Observability*, in contrast, refers to the ability to observe the effects of a specific internal, stimulated line of code or structure. Thus, a testbench generally offers limited observability, if it only observes what is on the external ports of the device or model. And all the internal signals and structures are often hidden from the testbench.

5.2.2 Types of traditional coverage metrics

A number of coverage metrics have been developed to determine the effectiveness and quality of the verification process. We summarize several coverage techniques in the following sections.

Ad-hoc metrics. Ad-hoc metrics include items such as bug rate, length of simulation after last bug found, and total simulation cycles. These metrics can provide interesting *quantitative* data, but when used for coverage metrics, they provide little *qualitative* data on how well the design has been verified or how much of the design has been left untested. The pressing question is: *If the bug rate reduces to zero while unverified features still exist, is the verification effort really complete?*

Programming code metrics. Most commercial coverage tools are based on a set of metrics originally developed for software program testing [Beizer 1990][Horgan et al. 1994]. These programming code metrics measure syntactical characteristics of the code due to execution stimuli. In other words, it is a measure of *controllability*. Examples are as follows:

- *Line coverage* measures the number of times a particular line of code was executed (or not) during a simulation.
- *Branch coverage* measures the number of times a section of code diverges into a unique flow.
- *Path coverage* measures the number of times a unique path through the code (including both statements and branches) is executed during a simulation.
- *Expression coverage* measures controllability of the individual variables, which contribute to the expression's output value.

For more on programming code coverage, see Drako and Cohen [1998] and Tasiran and Keutzer [2001].

observability and controllability related to programming code coverage

The value of code coverage is that it identifies *holes* (that is, something that has never been exercised during the course of verification). However, a shortcoming of programming code metrics is that they are limited to measuring the *controllability* aspect of our test stimuli applied to the RTL code. Activating an erroneous statement does not mean that the design bug would manifest itself at an observable point during the course of simulation.

Techniques have been proposed to measure the *observability* aspect of test stimuli by Devadas et al. [1996] and Fallah et al. [1998]. What is particularly interesting are the results presented by Fallah et al., which compare traditional line coverage and their observability coverage using both directed and random simulation. They found instances where the verification test stimuli achieved 100% line coverage, yet achieved only 77% observability coverage. Other instances achieved 90% line coverage, and achieved only 54% observability coverage.

functional correctness

Another drawback with programming code metrics is that they provide no qualitative insight into our testing for *functional correctness*. Kantrowitz and Noack [1996] propose a technique for functional coverage analysis that combines correctness checkers with coverage analysis techniques. In this chapter, we describe a similar technique that combines event monitors, assertion checkers, and coverage techniques into a methodology for validating *functional correctness* and measuring desirable events (that is, observable points of interest) during simulation.

In spite of these limitations, programming code metrics still provide a valuable, albeit crude, indication of which portions of the design have not been exercised. Keating and Bricaud [1999] recommend targeting 100% programming code coverage during block level verification. It is important to recognize, however, that achieving 100% programming code coverage does not translate into 100% observability (detection) of errors or 100% functional coverage. The cost and effort of achieving 100% programming code coverage must be weighed against the option of switching our focus to an alternative coverage metric (for example, measuring functional behavior using functional coverage specification).

State machine and arc coverage metrics . *State machine* and *arc coverage* is another measurement of controllability. These metrics address the number of visits to a unique state or arc transition as a result of the test stimuli. The value these metrics provide is in their ability to uncover unexercised arc transitions, which enables us to tune our verification strategy. Like programming code metrics, however, state machine and arc coverage metrics provide no measurement of observability (for example, an error resulting from arc transitions might not be detected), nor does it provide a measurement of the state machine's functional correctness (for example, valid sequences of state transitions).

Functional coverage metrics. User-defined functional coverage allows the designer, who possesses the greatest knowledge of the low-level design details and implementation assumptions, to specify functional coverage for points in the design that are known to be significant. The remainder of this chapter focuses on functional coverage.

5.2.3 What is functional coverage?

functional coverage versus programming code metrics

Functional coverage and programming code coverage tools offer different perspectives to coverage metrics. Code coverage tools take a blind approach to coverage by monitoring the design as a whole without specific knowledge of its operation. Conversely, since someone familiar with the design adds functional coverage, application domain knowledge is inherent.

functional coverage versus assertions

While RTL-implementation functional coverage and assertions can be implemented with the same form of specification, they focus on two different areas. To eliminate confusion between the two cases, this book refers to the detection and reporting of illegal behavior as *assertions*—and detection and reporting of expected (or desired) behavior as *functional coverage*. Essentially, assertions and functional coverage are both properties of the design. However, functional coverage provides indications of when a specific functionality of the design has been exercised. Nevertheless, both provide coverage feedback.

Definitions

Use the following definitions to guide your understanding of the terms we use in this book.

- *functional coverage* refers to the entire functional coverage methodology
- *functional coverage point* is a specific feature or event in the design that we want to monitor and include coverage information for in our functional coverage reports
- *functional coverage model* is a collection of functional coverage points that will be applied to a specific design
- *cross functional coverage* is an analysis of functional coverage over time

Grinwald et al. [1998] describe a coverage methodology that separates the coverage model definition from the coverage analysis tools. This enables the user to define unique coverage metrics for significant points within the design. They cite examples of user-defined coverage that targets the proper handling of interrupts and a branch unit pipe model of coverage. In general, user-defined functional coverage provides an excellent means for focusing and directing the verification effort on areas of specific concern.

internal monitors

For RTL implementation functional coverage, *internal monitors* are often used to capture functional coverage events. This provides an automated approach to obtaining coverage data. Since an engineer familiar with the design inserts the functional coverage monitor into the RTL, these monitors provide coverage feedback on areas of the design that the engineer feels are important.

5.2.4 Building functional coverage models

questions you should ask yourself related to functional coverage

Bergeron [2003] provides an interesting view of functional coverage. In his book, he describes the questions we must answer to build an effective functional coverage model.

- *What* should be covered? Questions such as *did the FIFO become full?* And *did I see a READ instruction?* are examples of what to cover.

- *Where* is the best place to monitor for the covered event? For example, when we have a READ instruction in a FIFO, we must ask ourselves if we should look for the READ instruction entering the FIFO or leaving the FIFO. Since there is a chance the instruction can enter the FIFO but never leave due to a reset condition, in some cases, it makes sense to look for the instruction leaving the FIFO. Monitoring the instruction entering the FIFO would give a false indication that the READ instruction was actually used.

- *When* should I look for the condition that I am covering? Is it on every clock edge or only on clock edges when the instruction valid signal is active?

- *Why* should we cover the event? Is the event interesting enough to monitor it, log it, include it in reports, and analyze it? Bergeron's example in this case is a 32-bit address decoder. Why would we want to monitor every 32-bit address when the decoder only analyzes the upper 4 bits to decode the address into 16 pages?

Asking these questions will aid you in understanding an effective process for applying functional coverage to your design.

shared ownership of functional coverage

The task of building functional coverage models is owned by both the verification and design engineers. The verification engineer specifies the high level aspects of the functional coverage model with a black-box view of the design. This occurs during the specification phase (refer to Section 1.4 "Phases of the design process" on page 14 for a description of the design process phases) and uses the specification as the primary input for specifying this portion of the functional coverage model. The verification engineer also describes coverage points down into the design at major interfaces between major blocks within the design during the architect/design phase. This approach takes a big picture view of the system. *(Are all processor transactions covered on the processor bus?)* Also during the architect/design phase, the design engineer specifies coverage points on the higher level design decisions, particularly at interfaces between the major blocks of the design. (*What is the protocol of the interface between these two major blocks?*) Finally, during the implementation phase, the design engineer takes a detailed picture view of the design. *(How is my FSM implemented?)* Additionally, the verification engineer adds coverage points to the testbench as it is implemented. Using this shared approach to functional coverage, a complete functional coverage model is created.

5.2.5 Sources of functional coverage

With this foundation understanding of functional coverage, we are ready to move forward and gain an understanding of the importance of how to build a functional coverage model. This section explores multiple sources that form the functional coverage model.

specification knowledge

Specification. The design specification is the initial source for specifying the functional coverage model. By building the functional coverage model utilizing the design specification, there is a direct mapping from the feedback generated by the functional coverage to the design specification. If this process of equating the functional coverage model to the design specification is done thoroughly, the test plan description becomes as simple as *reach 100% functional coverage*. This approach also follows the reconvergent model described by Burgeron [2003], because the functional coverage model written by the verification engineer is driven by the specification, not the implementation, which is driven by the design engineer.

verification knowledge

Test plan. One way to collect functional coverage is by creating and executing a test plan. Verification engineers create a test plan that details a list of required testing that must be accomplished to validate the design specification. Traditionally, the test plan is derived from the design specification. We then execute test plan items in a testing environment and verify the desired result. This verification step can be a manual or automated check, depending on the sophistication of the testing environment.

Writing specific directed tests to achieve functional coverage specified in the test plan is labor-intensive. Random test generation tools will often automatically cover many of the test plan item cases during random simulation. Functional coverage helps identify test plan items that are covered during random simulation.

design knowledge

Implementation. Specific design implementation knowledge is also captured by the functional coverage model. The engineer creating the design identifies interesting points within the design.

Functional coverage embedded in the RTL design produces an automated method of collecting functional coverage. Since the coverage points are associated with the model description, they are evaluated whenever the model is exercised, whether it be in block- or chip-level simulation environments. Functional coverage points can be as simple as *all input cases of a decoder have been seen* or as complicated as a multi-cycle handshake requirement of a bus specification.

Implementation-level functional coverage can be linked with test plans to more easily describe the individual test plan items and to provide a feedback path for when an individual test plan item has been completed. When functional coverage is combined with the detailed test plan, it answers the question: *Am I done verifying this design?*

assertion knowledge

Assertions. Assertions provide a form of functional coverage. While their objective is to detect and report illegal behavior, they also provide functional coverage when the input stimulus of the assertion has been activated but the condition being checked does not fail. For example, lets examine the following implication assertion:

```
assert always (A -> B);
```

This states that whenever A evaluates *true*, B must evaluate *true*. Notice that if A never occurs during the course of verification, we might get a false sense of security that the assertion is valid. Thus, an assertion is not very useful if the stimulus it is checking is never activated. Hence, triggering events associated with assertions are excellent functional coverage points. For example, the following PSL specification covers the triggering event in the assertion described above:

```
cover {A};
```

5.3 Does functional coverage really work?

The previous section explained what functional coverage is and how it can be used. This section provides concrete data to support the effectiveness of including functional coverage as part of your verification strategy.

5.3.1 Benefits of functional coverage

A summary of the major benefits of functional coverage is listed below. Many of these items are discussed in further detail elsewhere in this chapter.

- Functional coverage answers the question: *Am I done verifying the design?*
- Functional coverage provides feedback on areas of a design that the user designates as significant. This feedback gives an indication of how effective the project's current set of tests are in exercising the features of a design.
- Through analysis of the functional coverage results, verification teams optimize the tests that comprise the regression suites by removing tests that do not provide additional coverage. This optimization reduces the time and computer resources required for the project's regressions.
- Functional coverage provides specific feedback that can direct future verification efforts. For example, if the functional coverage shows certain features are not being tested, the design team can modify test generation algorithms to target those areas. Refer to Section 5.4.4, "Coverage analysis" and Section 5.4.6, "Coverage-driven test generation" for further details.
- Functional coverage provides the feedback needed to help determine the effectiveness of random environments and to help steer the configurations of these environments.
- Test plans can be written directly using property specification languages that generate concise and unambiguous functional coverage. Refer to Section 5.3.2, "Success stories" for more details.
- Functional coverage increases observability. Refer to Section 5.2.2, "Types of traditional coverage metrics" for more details on controllability and observability.
- A functional coverage methodology provides an automated means for collecting and reporting functional coverage.
- The impact of changes in the testing strategy can be seen through functional coverage over time.

5.3.2 Success stories

The following are examples of verification success achieved through functional coverage methodologies.

- Grinwald et al. [1998] describes how functional coverage allowed his team to trim the number of tests within their regression suite without a reduction in functional coverage. As a result, they significantly reduced the time and computer resources required to execute the regression suite.

- Ziv [2002b] describes two verification efforts that utilized functional coverage.
  - First, a PowerPC processor execution unit coverage model contained over 4400 functional coverage points. After 25,000 tests, about 64 percent of the functional coverage points were hit and additional tests were not providing a substantial increase in coverage. By analyzing the functional coverage, it was determined that two major areas were not well covered. By using this observation, adjustments to the test generators provided an immediate and substantial increase in coverage. Continued testing provided close to complete coverage of all functional coverage points.
  - Second, a branch unit for an S/390 processor coverage model contained about 1400 functional coverage points. By going through the process of adding the coverage points, the team was able to gain a better understanding of the design, even before beginning to collect coverage. In addition, the functional coverage provided information that identified several performance bugs.
- From the authors' own experience on the super scalable processor chipset sx1000 project at Hewlett-Packard, the coverage model was comprised of over 14,000 functional coverage points. By analyzing functional coverage for each verification environment, especially the random environments, the team found that several key test generation features they believed were enabled, were actually disabled. Additionally, functional coverage across all verification environments was merged and tracked across model releases. Analysis of the functional coverage results helped identify specific features that were not exercised in any verification environment. As a result, additional targeted tests were written to cover these features. Before tape out, 100% functional coverage was achieved. About 90% was easily attained through normal verification efforts. The last 10% required a substantial effort to reach.
- Bentley [2001] described the use of functional coverage on the Intel Pentium 4 microprocessor project. His team used almost 2.5 million unit-level coverage points combined with over 250,000 inter-unit coverage points to describe their coverage model. By the end of the project, the team was successful in covering 90% of the unit-level points and 75% of the inter-unit points, ultimately delivering high quality silicon.

5.3.3 Why is functional coverage not used

As with any technology, incorrect use produces useless results. And generally, people who encounter useless results avoid the technology. Functional coverage is no different. This section discusses some of the primary reasons given for avoiding functional coverage as part of an overall verification methodology. Following each point is a discussion of how a complete functional coverage methodology can eliminate the concern that is presented.

using functional coverage the "right" way is more an art than a science. [Ziv 2002b]

Too difficult. Some engineers consider the additional effort required to specify the functional coverage points too costly. It is true that adopting a new methodology involves a learning curve. And initially, the full benefits of the methodology are not necessarily realized. However, it is our experience that when a sound functional coverage methodology is adopted by the entire team, the rewards are far greater than the cost of implementation.

provides too much data

Too much data. Others suggest that functional coverage produces too much information to analyze. Engineers might confess they ignore the functional coverage reports because the data is not meaningful. If you are experiencing this scenario, make adjustments to your functional coverage methodology. Identifying functional coverage points is an important aspect of the process. If care is not taken to identify interesting coverage points, the functional coverage produced will be blurred by data that provides no significant feedback. However, by following a solid methodology that directs the team through the process of creating the functional coverage model, you can easily avoid this problem. Additionally, organizing the data in meaningful ways, as described later in this chapter, also eliminates problems with analyzing too much data.

provides too few results

Limited results. Some will argue that functional coverage provides a limited quantity of results. It is true that the quality and amount of results generated by functional coverage is directly related to the amount of effort spent identifying functional coverage points. If a limited effort is put forth to build a functional coverage model, the results will be limited. However, an effective functional coverage methodology (as discussed in Section 5.4) ensures that this docs not happen.

5.4 Functional coverage methodology

Just as with assertions, functional coverage is most effective when we institute a good functional coverage methodology. The following sections describe the basic practices that form a sound methodology. We take you from creating a process through visualizing the organization and devising the analysis involved in implementing your methodology.

5.4.1 Steps to functional coverage

how do I start?

One of the first obstacles to overcome is answering the question, *How do I start?* This section explores the basic steps involved in getting your functional coverage running.

choose the form of specification

The first step is to identify a form of functional coverage specification that you will use as the basis for your project (for example, PSL or SystemVerilog cover properties—or some other appropriate tool-specific form). Refer to Section 5.5, "Specifying functional coverage" for details.

start with the functional specification

With the functional coverage specification form in place, the next step is to begin the process of identifying and creating functional coverage points. The easiest place to start is with the functional specification for your system. Also refer to Section 5.4.2, "Correct coverage density" as a guide for further functional coverage points.

convert points to monitors

Once you have identified functional coverage points, convert them into monitors using the selected functional coverage form. The specifics will vary depending on the chosen form of specification. In some cases, the testbench tool or simulator is able to parse the functional coverage specification directly (for instance, PSL or SystemVerilog coverage constructs). In other cases, such as with a custom coverage technology or when working with tools that do not recognize languages such as PSL, additional work is required to translate the functional coverage specification into simulation monitors (see [FoCs 2003]).

collect the coverage data

A defined process must be in place to collect and merge the functional coverage results. Functional coverage is collected as soon as basic model stability is achieved. Even though only basic features are available at this point, functional coverage still provides effective feedback.

On super scalable processor chipset sx1000 project at Hewlett-Packard, an updated version of the design was released approximately once a week. Functional coverage was collected and posted on the project web site for each model release. It was the system test coordinator's responsibility to collect the results from all the different verification environments, merge them, and analyze how the results changed from release to release.

analyze the functional coverage

Once you collect functional coverage, you must analyze it. This analysis includes looking for improved coverage and noticing functional coverage points that were reached in previous releases but are no longer being hit. This data is used to make appropriate adjustments to the test generators and to write additional tests to increase the overall functional coverage. The practice of using the functional coverage to adjust the test generation can be a manual or automatic process. The concept of a test generator that automatically reacts to functional coverage is explored in Section 5.4.6, "Coverage-driven test generation".

5.4.2 Correct coverage density

To obtain the best information from functional coverage, the methodology must foster uniformity by including directions on where to add functional coverage points in the design. Without consistent placement, functional coverage has less impact. Assume that your analysis indicates that 90% of the functional coverage has been exercised. This data is not terribly meaningful if only 10% of the design is covered by functional coverage points. A well-defined guide to adding functional coverage points provides a measure of how much of the design is covered by functional coverage points.

Additionally, functional coverage is driven by an assessment of the features you want to validate. Achieving 100% functional coverage is an indication that the testing you want to do is complete. Ensuring the functional coverage model is a complete representation of what you wish to validate across the entire system is an important part of a successful functional coverage methodology.

This section explores some of the more common locations to apply functional coverage.

concise forms of specification

Table coverage. Teams often use tables to concisely describe the requirements for sections of a design. An error table may document the different detectable errors within a block and the different paths from which each error can be produced. Mapping functional coverage points to each entry in the table provides excellent feedback on how well the features in the table are being exercised.

corner cases

Interesting corner cases. Cover any interesting corner case that may be hit in a random environment. These cases are generally specific to each portion of the design. An engineer familiar with the design has knowledge of these corner cases. A corner case is any portion of the design that includes a complex algorithm or is an area of concern for the designer. One example would be to check for a specific sequence of requests to an arbiter that concerns the designer.

full and empty conditions

Queue levels. Add functional coverage points to capture how full a queue is (such as: full, half_full, full_minus_one, or full_minus_two). This provides better feedback on how much of a queue's depth is being utilized through simulations. Although it is difficult (and sometimes impossible) to actually get to a full state, feedback from functional coverage indicates if tweaks to test generation parameters or tests are moving the test in the right direction. Any resource that can become full falls into this category.

bypass and stall

Bypass and stall cases. Pipeline logic associated with registers often has a bypass or stall mode. Functional coverage monitors when these modes are used and these alternate paths are exercised. For instance, after a stall occurs, the next stage may need to avoid taking new data. In this case, functional coverage is used to capture the case when a stall occurred and the following data was available but held off.

Note the following caution: When the logic is idle, the bypass control could default to a specific case, which could cause the functional coverage condition to be active for a high percentage of the simulation if not properly qualified. This gives extraneous information. Always consider *when* the functional coverage point is of interest.

hw detectable errors

Hardware error detection cases. Often a design will include logic to record that certain hardware errors were detected. Place functional coverage points on hardware error log fields to capture an indication of how the error was reached. Since a hardware error can be detected in a variety of ways, ensure that all the error detection paths are identified in the coverage model. For instance, a particular detected error may have a feature that disables hardware error logging. Create functional coverage points to capture the cases in which the hardware error is enabled, detected, and logged as well as the cases in which the hardware error is disabled, detected, but not logged.

debug logic

Post-silicon debug logic. When a design includes specific logic to aid in post-silicon debug, this logic requires verification as well. Functional coverage can give feedback on the debug logic to ensure it has been tested.

internal sequences

Internal activity cases. Functional coverage can track the internal activities of the design in the same way it is used to track activities on external interfaces. Example of internal activities include scenarios such as *a packet was received with type=x* or *a recall response action was launched.* In these cases, use a different functional coverage point for each packet type or action. Go one step further and insert a different functional coverage point for each circumstance that results in a given action being launched.

traffic patterns

Traffic patterns. In addition to monitoring specific packets, it is often interesting to watch interesting traffic patterns. A traffic pattern is a sequence of events or transactions that are expected to occur within a system.

states

FSM states and state transitions. Most commercial coverage tools provide FSM state and arc (that is, transition) coverage. However, for multiple, interacting state machines spanning multiple hierarchies, these tools are not effective. High-level state machine activity is a great place to add functional coverage for important states or transitions.

use the spec

Specification-driven cases. Use the functional specification to identify useful functional coverage points. Include a mapping of these points back to the functional specification. For instance, include the specification heading number in the functional coverage point name. Shimizu and Dill [2002] extend this concept even further by automatically deriving testbench stimulus, checking properties, and functional coverage from the interface protocol specification.

what do I need to test

Other cases. Thinking about questions like: *What do I need to test in this section of the design*? Or *would you want to know if X occurred?* This is a good ways to come up with the functional coverage that should be added.

5.4.3 Incorrect coverage density

Use this section with great care to avoid over generalizing these concepts and misusing the guidance we offer. Although it is generally a good practice to use functional coverage in all areas of the design, there are some situations in which functional coverage does not add value to the coverage results. On the contrary, they produce a performance penalty and provide extraneous data that hinders coverage analysis.

too much

Too much data. As mentioned previously, one of the problems associated with any methodology is that you can generate more data than you can analyze in a reasonable amount of time. The goal is to obtain meaningful feedback from the functional coverage. If a functional coverage point does not provide meaningful feedback, it should not be a part of the coverage model. When you ensure that all the functional coverage provides meaningful information, you eliminate one of the reasons engineers avoid functional coverage.

always active

Functional coverage points that fire every cycle. In general, this type of functional coverage does not add enough information to the functional coverage to warrant the increased log file sizes. An alternative to eliminating this point is to constrain it to only activate at significant times.

too basic

Basic functionality. If the coverage point is too basic, the feedback is not useful, making the functional coverage reports so large that they become unmanageable (and eventually discarded). Associate functional coverage points with *interesting* cases that provide *real* feedback on how well the design is being exercised. When considering a coverage point, ask yourself what you will gain by knowing that the point was reached.

code coverage

Code coverage duplication. Functional coverage is superfluous if it provides feedback that is easily monitored by code coverage tools. If you need to trim back the number of functional coverage points, consider this area.

too many

Too many functional coverage points to add. If the set of functional coverage is too large to enumerate, use an alternative method to track coverage. An engineer should not define so many functional coverage points that it is impossible to write *and* track the functional coverage. This step prevents the coverage reports from becoming unwieldy. In certain cases, it may be possible to create a subset of the functional coverage points and use cross functional coverage to analyze the temporal relationship between those events.

5.4.4 Coverage analysis

An essential part of an effective functional coverage methodology is defining the process for analyzing and acting on data. This analysis occurs periodically throughout the project.

Coverage data organization

Functional coverage models produce a large volume of data. It is imperative that you have methods to sort and organize this data. This section explores methods that aid organization.

Taylor et al. [1998] describe one way to organize coverage data so that it can be used to effectively direct verification efforts. They divided the coverage analysis into the following four categories.

State transition. State transition analysis concentrates on complex state machines to ensure that all possible states and state transitions are exercised.

Sequence. Sequence analysis focuses on sequences of functional coverage over time. For instance, Taylor et al. [1998] used sequence analysis to ensure that every type of command leaving the CPU was followed by every type of command entering the CPU. This is equivalent to cross functional coverage.

Occurrence. The value of some functional coverage is that it was active at least once. The fact that it was activated many times gives no more coverage data than knowing that it was activated once. This type of functional coverage was analyzed with occurrence analysis. For instance, functional coverage can be used to ensure that all bits of an adder block have produced a carry-out.

Case. Case analysis deals with collecting statistics on the entire set of simulations. Statistics such as system configuration, instruction types issued, and bypass modes enabled are just a few examples of items from this category.

functional coverage groups

Another method for managing functional coverage better is *functional coverage grouping*. This provides a sorting parameter that allows functional coverage to be classified into functional categories, which is useful during data analysis. A functional coverage instance is associated with a unique coverage group. Some suggested coverage groups are listed below. However, groups are customized for each project's specific needs and include any number of categories that effectively organize a design's functional coverage.

Normal. Most designs include a *Normal* functional coverage group that consists of functional coverage points that don't have special functional characteristics. This is the default group for functional coverage points.

Not Reachable. A *Not Reachable* group includes functional coverage points that cannot be hit. While this group may seem superfluous, not reachable functional coverage points can appear in a design in cases where there are multiple instantiation of subblocks. For instance, a 1 to 4 decoder could have a functional coverage point associated with each decoded state. However, a particular instantiation of the decoder may only use half of the states. As a result, half of the functional coverage points associated with this instantiation will never be exercised due to the design implementation. If your functional coverage methodology does not allow for identifying these cases, the functional coverage reports will be difficult to interpret.

Error. Almost all designs will include logic for detecting, and in many cases correcting, runtime errors (for example, ECC logic and parity). An error group is useful because some tests in the regression suite will not target error cases. Since they are always identified as uncovered, when analyzing coverage of this subset of tests, the coverage data is skewed if the error monitors are included in the results. By including an error functional coverage group, you separate the error coverage data from other functional coverage or eliminate them from coverage reports for tests that don't target them.

Debug. Special logic is often added to debug features of a design, but (like error logic) they are not always targeted in normal testing. As with the *error* group, a *debug* group allows you to ignore any functional coverage associated with this logic when necessary.

Not Supported. Often during the design cycle, teams make trade-offs that remove a feature from the design. Any work completed on these portions of the design, including functional coverage points, is often left in the design for a future revision. However, these portions of the design will not be targeted in the verification efforts and any related functional coverage should be ignored. By placing these points in a *not supported* group, they are automatically excluded from functional coverage reports.

associating groups with functional coverage points

There are several methods to associate functional coverage points with a coverage group. A simple method is to prefix the name of each coverage point with the group name. By sorting the functional coverage points by name, the points are segregated into the individual groups. Another method, which requires additional infrastructure, is to use a mapping method. For instance, a separate file can list all the functional coverage points and their associated group. Example 5-1 shows a possible format for this file, which is parsed prior to generating reports to provide the group information.

Example 5-1 Associating functional coverage points with groups

```
FIFO_FULL                     QUEUE
FIFO_HALF_FULL                QUEUE
BAD_TRANSACTION               ERROR
```

Tracking functional coverage

Track functional coverage throughout the design and verification process, starting when the model is at an initial level of stability. When used early in the process, functional coverage ensures that design and tool features are accurately enabled. Later in the process, functional coverage ensures that all portions of the design are being exercised. You should track functional coverage along with each release of the model.

coverage across environments

It is important to track functional coverage across all the verification environments. Teams often use multiple verification environments to target specific portions of a design. By analyzing functional coverage data across all verification efforts, you avoid duplication. This also identifies areas of the design that have yet to be exercised by any environment.

tracking coverage over time

By tracking functional coverage across release models and over time, it is easy to watch progress in functional coverage. In addition, it is an accurate way to monitor the settings of the verification tools. For example, *feature X* is enabled, but one of the model releases included a bug that required the team to temporarily disable testing of *feature X*. After fixing the bug, it is important to re-enable the feature in the verification tools. However, this step is easily overlooked. By monitoring functional coverage, the team is alerted to the oversight, because events that occurred in previous model releases do not occur in the current model releases.

Actions to take

direct future verification efforts

At some point, the collected coverage will show a significant slow down in increasing coverage. This should prompt the team to make a change in test generators that results in an increase in the functional coverage. Also, use functional coverage to help direct the random verification environments, which are effective with corner cases of the design.

5.4.5 Coverage best practices

This section describes some practices that enhance a functional coverage methodology.

when to add functional coverage

Add during design creation. As described in Chapter 2, "Assertion Methodology" on page 21, for assertions, our experience shows that the best time to specify functional coverage is prior to and during RTL implementation. During this phase, the designer is most intimately familiar with the inner workings of the design and is aware of features that are the most interesting in coverage analysis. In addition, in the process of considering where to add functional coverage, designers often find design flaws (that is, before running verification tools). This methodology is also described by Foster et al. [2002] and Foster and Coelho [2001]

failed simulations

Exclude failed simulations. Since a failed simulation can be the result of the design stepping into illegal states, you can falsely reach functional coverage points. For this reason, do not collect functional coverage from simulations that failed.

names

Assign a name to every functional coverage point instantiation. Assigning a specific name makes it easier to track and sort the functional coverage. It is also useful to define a consistent naming convention. The following are two commonly used sorting conventions.

- Use the sub-block name as a prefix
- Use functional group names as prefixes or suffixes.

maximum reporting

Control the maximum number of times a functional coverage point can fire. As discussed with assertions, it is also useful to provide a mechanism to cap the maximum number of times a functional coverage point will fire. This is especially important if the coverage data is used only to see if certain portions of a design have been exercised (occurrence coverage). This practice reduces the size of functional coverage logs, which saves disk space and improves simulation performance. Note: Since all functional coverage data is not logged, if you use this feature, results from cross functional coverage are incomplete.

when did I reach it

Watch for specific functional coverage points. In volume simulations, capturing functional coverage is an automated process. However, some functional coverage points could be very difficult to hit. For these functional coverage points, it is important to capture the simulation parameters, such as test name or random seeds, that were used in the simulation that hit them. To address this need, implement a method to query the functional coverage log from each simulation looking for any *hard to reach* functional coverage points. Another option is to use the severity level of functional coverage to force the simulation to have the appearance of a failure when the points are reached. If they fire, the process archives the simulation parameters along with the functional coverage. With this information, you can recreate and analyze the simulation in further detail.

common logfile format

Use a common logfile format. An important part of functional coverage is to define a common log file format that you will use across the entire project. With a common format, whether it is a database or a text file, you can merge and compare the coverage data in many ways, including between model releases and across verification environments.

what to log

Record the right information. The important pieces of information to record include the following:

- Functional coverage point name
- Indication of instantiation (this could be the hardware path or other identifying information)

- Time (time is only important if a cross functional coverage analysis is performed)
- Coverage group for sorting the data (if needed).

multiple instances

A typical design has multiple instantiations of a module, so it follows that you have multiple instances of a functional coverage point with a single name. To effectively identify each instance, the functional coverage technology must log the hardware path to each functional coverage point.

merge data

Merge data across simulations. For volume progression simulations, a method is required to merge functional coverage from all simulations. This provides access to all the data that would be needed for any type of functional coverage analysis, including cross functional coverage. However, saving all this data requires a large amount of storage. Alternatively, if you do not intend to use cross functional coverage, configure the reporting mechanism such that it does not log the time information.

A variety of statistics can be generated, such as, average number of functional coverage point firings per simulation as well as maximum and minimum number of functional coverage point firings across all simulations. Examples of functional coverage reports are described by Kantrowitz and Noack [1996] and in Example 5-2.

Example 5-2 Functional coverage report

```
Group        Id                  Total   TestsHit    Avg      Max      Min
-------------------------------------------------------------------------
Normal;BLK  QMU_XX4_FROM_1       119528  1787        67       270      0
Normal;BLK  QMU_XX4_FROM_2       162409  1946        83       463      0
Normal;BLK  QMU_XX4_BYPASS       0       0           0        0        0
<...>
NoReach;BLKQMU_PR4_FROM_XIN      0       0           0        0        0
<...>

There were 462 out of 474 total Normal points hit in BLK (97.5%)
There were 0 out of 25 total NoReach points hit in BLK (0.0%)
```

regression testing

Regression and progression testing. The early stages of the verification process focus on *regression testing*. Regression testing involves running a set of known, working tests (called a regression suite) on each successive model release to ensure that new features do not create new bugs in previously verified portions of the design. In simpler terms, all tests that had been working continue to work. Regression testing is backward-looking. It does not concentrate on new features. It concentrates on previously verified features. A good regression suite evolves over the life of the project. As more tests are created and validated, they are added to the regression suite.

Functional coverage is a critical part of regression testing. By capturing and analyzing functional coverage from regression testing, you assure your team that the functional coverage is, at the very least, staying constant. In other words, the regression suite should continue to produce the same level of functional coverage for each model release.

progression testing

Progression testing is forward-looking. It explores new regions of the design and tries to exercise existing areas of the design in different ways than in the past. The purpose of *progression testing* is to increase the total number of cycles simulated in hopes of finding hidden corner case bugs. The exact role of progression testing is somewhat dependent on where you are in the verification life cycle.

Progression testing is used early in the verification life cycle, especially in the random environments, to gain additional coverage and to exercise hard-to-reach corner cases. Functional coverage plays a valuable role by reporting how much of the design is being exercised. In this phase, capture and analyze functional coverage to ensure that the effort you expend is really increasing coverage.

Later in the verification life cycle, when the error rate is minimal and functional coverage has reached targeted levels, the emphasis of progression testing shifts to focus on increasing the total number of simulation cycles in hopes of finding those remaining corner case bugs. In this phase of progression testing, functional coverage data has a reduced value and the simulation performance is the most important factor. In this progression testing phase, it is acceptable to turn off functional coverage to improve simulation performance.

correctness

Functional coverage correctness. If functional coverage is to provide the intended coverage feedback, teams must correctly specify the coverage points. Several possibilities exist for ensuring the correctness of the functional coverage points implementations.

- **Directed tests.** One method to ensure that functional coverage is specified correctly is to write a simple directed test that stimulates the model in such a way that the functional coverage point in question fires. Unfortunately, this method is time consuming. Thus, you cannot use this method exclusively if your design has a large number of functional coverage points.

- **Linting**. Use lint checkers not only for the synthesizable portion of a design, but also for the functional coverage points. Simple errors in the coverage specification are easily caught using this method.

- **Peer reviews**. Peer reviews are just as effective for finding problems with functional coverage specification as they are for finding errors in the actual design implementation. Review the coverage specification as part of normally scheduled design reviews or at reviews that specifically focus on functional coverage.

- **Validation checks using simulations**. An effective way to verify the correctness of functional coverage specification is to compare the coverage data obtained from simulations with the stimulus for that simulation. Using deductive reasoning skills, you can determine whether the scenarios required to generate the functional coverage were, in fact, present. If they were not, analyze the coverage points for errors. Likewise, investigate coverage points that do not fire even though it is reasonable to expect that they would.

5.4.6 Coverage-driven test generation

reactive testbench

Random test generators are giving verification engineers a new approach for achieving their objectives. As the use of random generators increases, the next step to increased simulation productivity is to combine the observability provided by functional coverage with the controllability provided by random test generators. By including feedback from functional coverage into the random test generators, the focus of the generators can be shifted based on functional coverage levels or as the result of reaching a specific functional coverage point. For example, a random generator may include a method to inject a full bandwidth stream of transactions in hopes of filling queues. By providing feedback that the queues of interest have been filled, the test generator can immediately switch to another focus, instcad of involving the entire simulator in keeping the queue full. This process of feeding back functional coverage to test generators is often called a *reactive testbench.*

Another tool available to the verification engineer is Hardware Verification Languages (HVLs). HVLs provide a language specifically designed for testbench generation. In addition, most modern HVLs provide mechanisms for creating reactive testbenches.

Reactive testbench generation requires application-specific knowledge of your design. Methods for easily generating reactive testbenches have recently been investigated , [Adir et al., 2002a], [Adir et al., 2002b], [Benjamin et al., 1999], [Geist et al.., 1996], [Ziv et al., 2001]. In some cases [Ur and Yadin, 1999], a separate description language was developed to describe a parallel model of the system by which coverage feedback was provided. They found that the cost associated with developing a reactive testbench was less than the effort required to manually provide the feedback. With emerging hardware verification languages (HVL), this feedback path can be captured along with the testbench.

5.5 Specifying functional coverage

There are many excellent tool-dependant and hardware verification language (HVL) solutions available for measuring functional coverage (for example, *e* [e Language Reference Manual] or *OpenVera* [OpenVera Language Reference Manual])—and we recommend you take advantage of these offerings whenever possible to reduce project resources in developing a functional coverage model. As previously stated, often these tool-dependant solutions focus on higher-level forms of functional coverage, such as transactions. Hence, methodologies and convenience around RTL implementation functional coverage is rarely addressed. When creating your own *RTL implementation* functional coverage methodology, we recommend a tool-independent solution that can be delivered with IP or re-usable blocks. This allows you to leverage RTL functional coverage across all verification environments and tools.

specify once

Functional coverage points should only have to be *specified once* in the design and then supported by all verification environments and tools. Verilog-based assertion solutions such as OVL and SystemVerilog generally enable this by default, as the functional coverage points are evaluated by the Verilog simulator. For property languages such as PSL, ensure that the tool suite your project uses supports the language.

5.5.1 Embedded in the RTL

Bening and Foster [2001] describe a simple method to implement functional coverage by embedding the functional coverage points directly into the Verilog RTL. Example 5-1 illustrates functional coverage that detects when a queue reaches its full mark.

Example 5-3 RTL functional coverage implementation

```
`ifdef COVERAGE_ON
// look for a queue-full condition
always @(posedge clk) begin
   if (reset_n == 1'b1 && q_full) begin
         $display("COV_Q_FULL @ %0d:%t:%m", $time);
   end // end if (...q_full...)
end
`endif
```

While this approach provides an easy method to add functional coverage, you must type a substantial amount of text for each functional coverage point that you add. This may be a deterrent for some designers. Also, since the output is through a system function, there is limited control of the messaging. Ensure that you use a common message format for all instantiations using this method. Since output is to standard output in this example, a method is required to extract the functional coverage messages from messages produced by other sources. In this example, a prefix of "COV_" identifies functional coverage messages.

5.5.2 Functional coverage libraries

In Chapter 3, "Specifying RTL Properties" on page 57, we discussed the specification for a library of assertions through the OVL. Although the OVL does not currently provide a direct way to specify functional coverage, use this concept to create your own set of reusable templates that specify functional coverage points and encapsulate the functional coverage point's detection and reporting methods in an RTL module. This simplifies what an engineer must type when instantiating functional coverage points. In addition, it provides a centralized location for controlling, maintaining, and optimizing the detection and reporting mechanisms. This approach also adds an abstraction layer between the designer and the assertion language. The language used within the template library can be moved to the latest technology without the need to change the instantiated coverage points. Example 5-4 and Example 5-5 illustrate this method for a *queue full* functional coverage point

Example 5-4 Coverage module in template library (Verilog)

```
module cover_monitor (clk, reset_n, test);
input clk, reset_n, test;
parameter event_id="COV_";
  `ifdef COVERAGE_ON
  // look for test condition
  always @(posedge clk) begin
     if (reset_n == 1'b1 && test) begin
        $display("%s @ %0d:%t:%m", event_id, $time);
     end // end if (...test...)
  end
  `endif
endmodule
```

Example 5-5 Coverage module instantiation (Verilog)

```
// detect a queue full condition
cover_monitor #("COV_Q1_FULL") dv_q_full(clk, reset_n, q1_full);
```

5.5.3 Assertion-based methods

You can use any assertion solution that allows for severity levels as the basis for functional coverage. By setting the severity for functional coverage instantiations to a non-error level, assertions can become functional coverage points.

Open Verification Library (OVL)

The Accellera OVL (www.openveriflib.org) provides a library of assertions that include severity levels and reporting parameters. These allow you to convert the assertions to functional coverage points and customize reporting. With this mechanism, the severity level changes the reporting information and the action required when the functional coverage point is activated.

Example 5-6 shows how the OVL **ovl_error** task can be modified to incorporate a separate severity level so that the OVL modules can be used for functional coverage. Example 5-7 shows how you can then use the OVL to define a functional coverage point. Please note when using this method, the event condition must be specified in a negative sense because assertions are activated when the condition fails.

Example 5-6 Modifying ovl_error for functional coverage

```
#define COVERAGE 2
task ovl_error;
    input [8*63:0] err_msg;
  begin
    if (severity_level != 'COVERAGE) begin
      error_count = error_count + 1;
      `ifdef ASSERT_MAX_REPORT_ERROR
      if (error_count <=`ASSERT_MAX_REPORT_ERROR)
      `endif
          $display("OVL_ERROR :%s:%s:%0s: severity %0d : time %0t:%m",
                   assert_name, msg, err_msg,severity_level, $time);
      if (severity_level == 0) ovl_finish;
    endif
    else
      $display("OVL_COV :%s:%s:%0s : severity %0d : time %0t:%m",
                   assert_name, msg, err_msg,severity_level, $time);
  end
endtask
```

Example 5-7 Using OVL for functional coverage

```
assert_always #('COVERAGE,0,"Q_FULL") myQfull (clk, reset_n, !q_full);
```

SystemVerilog

The current SystemVerilog specification from Accellera (www.accellera.org) includes a language extension for adding assertions to the Verilog language. While often used for assertions, this new language extension is also effective for functional coverage points when you use appropriate severity levels. For example, use a severity level of *Error* for assertions and a severity level of *Info* for functional coverage points. At the conclusion of each simulation, use a post processing step to filter out messages that were printed with severity level *Info* and log or analyze them as you would any other style of functional coverage. Additionally, SystemVerilog is currently defining a `cover` feature that directly supports functional coverage.

As SystemVerilog is finalized, adopted as a new standard, and implemented in the various vendor simulators, it is possible that some of the differentiating features of the various tools will be to provide automatic logging of SystemVerilog assertions. Furthermore, it is hoped that vendors will include analysis tools to aid in reviewing functional coverage.

Example 5-8 shows how to use SystemVerilog to define a functional coverage point. Please note that the exact syntax may change when SystemVerilog specification is finalized.

Example 5-8 Using SystemVerilog for functional coverage

```
always @(posedge clk) begin
  if (reset_n)
    myQfull: cover (q_full) $info("queue was full");
end
```

PSL The current formal property language specification from Accellera is named Property Specification Language (PSL). While PSL has only recently been finalized by Accellera, it is already beginning to be supported by simulators. Example 5-9 shows how to use PSL to define a functional coverage point.

Example 5-9 Using PSL for functional coverage

```
default clock = (posedge clk);
sequence qFullCondition = {reset_n ? (q_full) : 1'b0};
cover qFullCondition;
```

5.5.4 Post processing

Another implementation of functional coverage includes a post-processing mechanism. This can be used exclusively to generate functional coverage by processing signal logs. Alternatively, you can use it to perform cross functional coverage analysis. In either case, the only simulation-time logging required is to capture the signal states you will need to evaluate during the post-processing steps. Once the simulation is complete, a custom post-processing script can parse the signal logs looking for functional coverage cases. This process was used by Kantrowitz and Noack [1996].

5.5.5 PLI logging and reporting

Standard Verilog-based functional coverage reporting techniques discussed earlier in this section are limited by the reporting capabilities of the language itself—since they are built around the system task `$display`. Since `$display` reports to `standard output`, this normally requires some sort of post processing step to parse a log of output. An alternative to this approach is to develop custom report libraries through the PLI interface. Since this ties the coverage points into a custom program, the possibilities are limitless for the types of processing that can be done.

5.5.6 Simulation control

The concepts of controlling functional coverage are similar to those of controlling assertions (discussed in Section 2.4 "Assertions and simulation" on page 42). Bening and Foster [2001] also describe several elements of controlling functional coverage. The exact method you use will vary depending on your functional coverage technology.

Enabling functional coverage. Since it includes additional checks and reporting, functional coverage reduces the performance of a simulation. For this reason, it is important to use some mechanism to enable and disable functional coverage. Refer to Section 5.4.5, "Coverage best practices" for more details on the need to control functional coverage.

'ifdef

A simple mechanism such as `'ifdef COVERAGE_ON` provides a coarse process for controlling functional coverage. This method requires that *all* functional coverage be enabled or disabled.

global enable signal

A second method includes a *global enable signal* within the functional coverage point. The global enable signal ensures the coverage monitoring does not begin until after the system is out of reset and possibly initialization. This method is used within the OVL template libraries as shown in Example 5-10. The testbench drives the signal `'TOP.dv_coverage_enable` shown in this example.

Example 5-10 Using global enable to control functional coverage (Verilog)

```
// check to ensure reset & init is done
if ('TOP.dv_coverage_enable) begin
  always @(posedge clk) begin
    if (reset_n == 1'b1 && q_full) begin
      $display("COV_Q_FULL @ %0d:%t:%m", $time);
    end // end if (...q_full...)
  end // end always (...)
end // end if ('TOP...)
```

individual enable signal

Finally, use an additional *enable signal* to the port list of the functional coverage module itself. This method, while similar to the previous method, adds the capability of grouping functional coverage into categories. Each group of functional coverage within a single category uses a separate enable signal. This method gives finer control over enabling and disabling functional coverage.

Reset and initialization. In general, most functional coverage data is only useful when the design is out of reset and has been initialized. For this reason, it is important to provide reset and initialization state as inputs to the functional coverage points.

5.6 Functional coverage examples

Consider the FIFO model described in Example 3-6 (see page 66). The following examples show relevant functional coverage points we could implement for this FIFO.

Example 5-11 ***OVL* FIFO coverage model**

```
// when checking functional coverage with an OVL, the expression
// must be expressed in the negative.

`ifdef `COVERAGE_ON
// FIFO full
assert_always #(`COVERAGE,0,"FIFO_FULL") myFIFOFull (clk, reset_n,
            !({push,pop}==2'b10 && cnt==FIFO_depth-2));

// FIFO full-1
assert_always #(`COVERAGE,0,"FIFO_FULL_M1") myFIFOFullM1 (clk,
            reset_n, !({push,pop}==2'b10 && cnt==FIFO_depth-3));

// FIFO full-2
assert_always #(`COVERAGE,0,"FIFO_FULL_M2") myFIFOFullM2 (clk,
            reset_n, ({push,pop}==2'b10 && cnt==FIFO_depth-4));

// FIFO empty
assert_always #(`COVERAGE,0,"FIFO_EMPTY") myFIFOEmpty (clk,
                reset_n, !({push,pop}==2'b01 && cnt==1));

// unneccessary coverage point
assert_always #(`COVERAGE,0,"FIFO_EMPTY") myFIFOPush (clk,
                reset_n, !(push==1'b1));
`endif // COVERAGE_ON
```

Example 5-11 shows the coverage model for the FIFO. The obvious points of the FIFO being full and empty are covered. Notice in these points how the event condition provided to the monitor includes both the `cnt` level and the strobe indicating a push or a pop. If we consider only the `cnt` value, the coverage point would fire every clock cycle that the FIFO was full or empty. In many cases, the FIFO may become full and stay full for five cycles before a value is popped out of the FIFO. We want the functional coverage to indicate we filled the FIFO one time in this case, not five times. Next, we include two additional coverage points that indicate when the FIFO is one entry short of full and two entries short of full. (When an engineer is working on a test to fill the FIFO, it is helpful to have some feedback on whether the test is getting close to the target of filling the FIFO.)

We discussed how important it is to carefully consider whether to add a specific functional coverage point. In Example 5-11, an additional functional coverage point monitors the number of times a value was pushed onto the FIFO. While this is a valid coverage point, we should exclude it because it does not provide valuable information. Example 5-12 and Example 5-13 show the same functional coverage model implemented in PSL and SystemVerilog, respectively. In these two examples, we do not include the unnecessary functional coverage point.

Example 5-12 *PSL* FIFO coverage model

```
default clock = (posedge clk);

sequence COVER_FIFO_FULL ={reset_n && rose(cnt==(FIFO_depth-1))};
cover COVER_FIFO_FULL;

sequence COVER_FIFO_FULL_M1 = {reset_n && rose(cnt==(FIFO_depth-2))};
cover COVER_FIFO_FULL_M1;

sequence COVER_FIFO_FULL_M2 = {reset_n && rose(cnt==(FIFO_depth-3))};
cover COVER_FIFO_FULL_M2;

sequence COVER_FIFO_EMPTY = {reset_n && rose(cnt == 0)};
cover COVER_FIFO_EMPTY;
```

Example 5-13 also shows an alternative method to capture only the initial entry into a full or empty condition.

Example 5-13 *SystemVerilog* FIFO coverage model

```
`ifdef `COVERAGE_ON
always @(posedge clk) begin
  if (reset_n) begin
    // FIFO full
    myQfull: cover property ($rose(cnt == (FIFO_depth-1)));

    // FIFO full-1
    myQfullm1: cover property ($rose(cnt == (FIFO_depth-2)));

    // FIFO full-2
    myQfullm2: cover property ($rose(cnt == (FIFO_depth-3)));

    // FIFO empty
    myQempty: cover property ($rose(cnt == 0));
  end
end
`endif // COVERAGE_ON
```

Continue to consider the FIFO example. If the design uses this FIFO to store command packets prior to a decode block, it is interesting to know when the different command packets are popped from the FIFO by the decode block. Example 5-14 shows examples of functional coverage points that monitor for the following command packets: `` `READ ``, `` `WRITE ``, `` `IO_READ ``, `` `IO_WRITE ``.

Example 5-14 *OVL* **FIFO command packet coverage points**

```
// when checking functional coverage with an OVL, the expression
// must be expressed in the negative.

`ifdef `COVERAGE_ON

assert_always #(`COVERAGE,0,"CMD_READ") read (clk, reset_n,
          !(pop==1'b1 && data_out==`READ));

assert_always #(`COVERAGE,0,"CMD_IO_READ") io_read (clk, reset_n,
          !(pop==1'b1 && data_out==`IO_READ));

assert_always #(`COVERAGE,0,"CMD_WRITE") write (clk, reset_n,
          !(pop==1'b1 && data_out==`WRITE));

assert_always #(`COVERAGE,0,"CMD_IO_WRITE") io_write (clk, reset_n,
          !(pop==1'b1 && data_out==`IO_WRITE));
`endif // COVERAGE_ON
```

5.7 AHB example

In this section, we discuss a transaction modeling technique proposed by Marschner et al. [2002] for the Advanced High-Performance Bus (AHB) protocol—supported by the ARM Advanced Microcontroller Bus Architecture (AMBA). The technique they proposed is based on specifying a set of PSL sequences, which represent various AHB transactions. The set of sequences are then combined into an assertion that requires only valid transactions to occur following the completion of any previous transaction (indicated by `hready` going high), as demonstrated in Example 5-15.

Example 5-15 *PSL* **AHB valid transactions following the completion of any previous transaction**

```
assert always {hready} |=> { SERE_AHB_BURST_MODE_READ
                           | SERE_AHB_BURST_MODE_WRITE
                           | SERE_AHB_SINGLE_READ
                           | SERE_AHB_SINGLE_WRITE
                           | SERE_AHB_INACTIVE
                           | SERE_AHB_RESET
                           };
```

AHB is a pipelined bus with all transfers taking at least two cycles to complete. For example, consider two transfers A and B demonstrated in Figure 5-1:

Figure 5-1 **AHB pipeline bus**

< A's addr phase >   < B's addr phase >

< A's data phase >   < B's data phase >

The Slave's response to the address (and control) phase occurs one cycle later in the data phase. The Slave can either set `hready` high, to acknowledge the data, or set it low inserting a wait cycle in the data phase (and consequently, the next address phase).

The address phase also includes control. Hence, an `htrans` transfer occurs in the address phase and the slave response (with `hready` & `hresp`) occurs one cycle later.

Marschner et al. [2002] demonstrated how to specify the AHB burst-mode read transaction assertion as a set of sequences. We now extend their transaction specification discussion by demonstrating how to create a functional coverage model for an AHB burst-mode read.

A burst-mode read transaction can be specified as a sequence of a *first read* operations, followed by one or more sequences of *next read* operations, for which the address and data are pipelined. Hence, a simplified description for the burst-mode read transaction can be described as follows. The transaction initially begin when the slave is operating on the previous data transfer. At this point, the master will request a first data transfer. Then, for all subsequent burst-mode read transfers (until the transaction completes), the slave will operate on the previous data transfer while the master is requesting the next data transfer or it signals that it is busy (that is, `hburst=='BUSY).`

Example 5-16 demonstrates how to specify the transaction for a burst-mode read as a sequence in PSL. This sequence can then be used to build a functional coverage model (using the PSL `cover` directive). During the course of simulation, any occurrence of a burst-mode read transaction is reported. Conversely, if burst-mode read transaction is never detected across our entire suite of regression tests, then we know that some fundamental aspect of the design has gone untested. Note that this specification, although useful for measuring functional coverage, would have to be made a little more precise when used as an assertion (for example, an assertion would need to account for a BUSY transfer).

Example 5-16 *PSL* AHB read burst mode transaction

```
'define AHB_WAIT (!hready && (hresp=='OKAY))
'define AHB_OKAY (hready && (hresp=='OKAY))
'define AHB_ERROR (hresp=='ERROR)
'define AHB_SPLIT (hresp=='SPLIT)
'define AHB_RETRY (hresp=='RETRY)

sequence SERE_AHB_SLAVE_RESPONSE = {
  'AHB_WAIT[*];
  {
    { 'AHB_OKAY}
  | { {!hready;hready} && {'AHB_ERROR [*2]} }
  | { {!hready;hready} && {'AHB_SPLIT [*2]} }
  | { {!hready;hready} && {'AHB_RETRY [*2]} }
  }
};

`define AHB_FIRST_TRANS (htrans==`NONSEQ)
`define AHB_NEXT_TRANS (htrans==`SEQ)
`define AHB_MASTER_BUSY (htrans==`BUSY)
`define AHB_READ_INCR(!hwrite && (hburst==`INCR))

// slave response to the previous data in parallel with the master's
// assertion of the control signals for the next address

sequence SERE_AHB_READ_FIRST = {
  {SERE_AHB_SLAVE_RESPONSE} &&
  {('AHB_FIRST_TRANS && 'AHB_READ_INCR)[*]}
};

sequence SERE_AHB_READ_NEXT = {
  {SERE_AHB_SLAVE_RESPONSE} &&
  {
    {('AHB_NEXT_TRANS && 'AHB_READ_INCR)[*]} |
    {'AHB_MASTER_BUSY[*]}
  }
};

sequence SERE_AHB_BURST_MODE_READ = {
  {SERE_AHB_READ_FIRST}; {SERE_AHB_READ_NEXT}[*]
};

cover {SERE_AHB_BURST_MODE_READ};
```

Note that in addition to specifying the large transaction as a functional coverage point, we recommend that you also specify the various sequence segments to provide finer granularity in identifying exactly what behaviors have been covered.

5.8 Summary

This chapter introduced the concept of functional coverage and discussed its role in the verification process. We provided specific details to allow the reader to construct an effective functional coverage methodology. Finally, we discussed sources of functional coverage technology.

CHAPTER

6

ASSERTION PATTERNS

Patterns, which has emerged as a popular topic of discussion within the contemporary software design community, is a convenient medium for documenting and communicating design insight as well as design decisions (for example, design assumptions, structures, dynamics, and consequences). The origin of this notion is actually rooted in contemporary architecture (that is, the design of buildings and urban planning [Alexander 1979]). However, their descriptive problem-solving form also makes patterns applicable (and useful) across the broad and varied field of engineering. In this chapter, we introduce patterns as a system for documenting (in a consistent form) and describing commonly occurring assertions found in today's RTL designs. We propose a pattern format that is ideal as a quick reference for various classes of assertions, and throughout the remainder of the book we use it in our assertion descriptions. In addition, the format we propose is useful when documenting your own assertion patterns and increases their worth when they are shared among multiple stakeholders.

6.1 Introduction to patterns

A *pattern*, by definition, is an observable characteristic that recurs. Furthermore, it often serves as a form or model proposal for imitation.

In *Software Patterns*, James Coplien [2000] describes a pattern as follows:

> *I like to relate this definition to dress patterns. I could tell you how to make a dress by specifying the route of a scissors through a piece of cloth in terms of angles and lengths of cut. Or, I could give you a pattern. Reading the specification, you would have no idea what was being built or if you had built the*

> *right thing when you are finished. The pattern foreshadows the product: it is the rule for making the thing, but it is also, in many respects, the thing itself.*

A succinct and intuitive definition of a *pattern*, in the context of conveying design intent, was provided by Appleton [2000]:

Pattern definition

> *A pattern is a named nugget of insight that conveys the essence of a proven solution to a recurring problem within a certain context amidst competing concerns.*

Appleton goes on to say:

> *A pattern involves a general description of a recurring solution to a recurring problem replete with various goals and constraints. But a pattern does more than just identify a solution, it explains why the solution is needed!*

Hence, you will see in the next section that a pattern format (sometimes referred to as a pattern language) provides a systematic and consistent method of documentation describing a *recurring solution* to a *recurring problem*.

6.1.1 What are assertion patterns?

Applying pattern-based approaches to present, codify, and reuse property specification for finite-state verification was originally proposed by Dwyer et al. [1998]. In this section, we build on the Dwyer discussion by categorizing the assertion patterns related to the three distinct phases of design, *specification*, *architect/design*, and *RTL implementation,* introduced in section 1.4 "Phases of the design process" on page 14.

Property structure and rationale

The goal in creating an *assertion pattern* is to present two interdependent components (that is, *property structure* and *rationale*) about a particular characteristic or verification concern associated with a particular design. The property structure serves as a model for possible implementations, while the rationale tells you under what conditions the property should be used and examines the various trade-offs and variations. Their interdependence is essential, for without a property structure, the assertion pattern's rationale is superficial or meaningless, and without a rationale, the assertion pattern's property structure is perplexing and of little use.

Pattern categories. Just as specifying assertions can range from high-level system properties and block-level interfaces down to lower-level RTL implementation concerns—assertion patterns can be categorized in a similar fashion [Riehle and Zullighoven 1996].

Categories of assertion patterns

- *Conceptual patterns* express higher-level global or system-level properties that concern large-scale components of an application domain.
- *Design patterns* are a refinement of the higher-level architectural patterns into medium-scale subsystem properties. Typically, these patterns describe block-level component relationships and their interfaces.
- *Programming patterns* are lower-level patterns specific to the RTL implementation. These patterns describe how to implement particular aspects or details of a component using the features of a given hardware description language.

Notice that these categories map into the three distinct phases of the design process, discussed in section 1.4 "Phases of the design process" on page 14. In other words, defining *conceptual patterns* is appropriate during the *specification phase*, while *design patterns* should be defined during the *architect/design phase*, and similarly *programming patterns* must be considered during the *RTL implementation phase.*

6.1.2 Elements of an assertion pattern

Problem, solution, and context

Fundamental to a pattern is the format used to describe (document) the *solution* to a *problem* in a *context*. A number of pattern formats have been proposed, such as the *Alexander form* [Alexander 1977] and the *Gang of Four form* [Gamma et al. 1995]. Coplien [2000] provides a comprehensive survey of various proposed forms as well as a detailed description of the elements contained within a pattern form. Many pattern experts (and critics) will argue the advantages of one format over another. For our purposes, consistency in documenting and conveying the assertion intent is the primary goal. Hence, the pattern format we propose for assertions draws from the various formats to suit our needs. The pattern elements (or sections) we recommend are:

Recommended pattern elements

Pattern name. This is important since it identifies an assertion solution and quickly becomes a part of the design team's vocabulary, which aids communication between engineers. We recommend a meaningful pattern name that is either a single word or short phrase.

Problem. Describe the problem to be solved. We recommend a concise statement, which helps engineers decide if this particular problem is applicable to their own (that is, whether to read further).

Motivation. Describe a scenario that illustrates the design problem.

Context. Describe (in a broader sense than the motivation section) the situations in which the problem recurs, and to which the solution applies.

Solution. Provide the details used to solve the stated problem. We recommend a solution that is detailed enough for the reader to know what to do, but general enough to be applied to a broader class of similar problems.

Considerations. This section identifies caveats for usage or suggests alternative patterns for certain situations. In addition, this section recommends alternative applications for the pattern or novel usage.

Applying patterns

The remainder of this chapter discusses a number of common assertion patterns found in a typical RTL design. Hopefully, by reading through these examples, you will see that the power of this pattern format is its ability to clearly describe assertions. We also expect that the assertion patterns will aid in application of assertions for your logic.

6.2 Signal patterns

In this section we define a set of patterns related to signal use within an RTL model. One pattern that may not immediately come to the designer's mind, but does immediately affect design operation, is undriven inputs that evaluate to Z or signals derived from unconnected ports that evaluate to X. Other patterns related to signals include multi-bit range checks as well as one-hot checks and gray codes.

6.2.1 X detection pattern

Pattern name. X detection

Problem. Detect unconnected ports and undriven signals, as well as X assignment propagation.

Motivation. During RTL development (that is, initial coding or code modifications and edits) the engineer often leaves an unconnected input port to a module, defines a new variable without an assignment, or neglects to drive a signal within a testbench. The *X detection* pattern is useful for identifying and isolating this class of problem.

Context. In addition to unconnected signals, RTL modeling that contains X assignment and detection is problematic and should be avoided [Bening and Foster 2001].[1] Problems typically encountered include missing functional bugs associated with startup (that is, during the reset process) as well as reduced performance of the RTL simulation model. Nonetheless, the development of today's complex chips often involves multiple designers and verification engineers and a significant amount of design reuse (that is, internally- and externally-developed IP). Hence, IP consumers often have little or no control over the coding of the RTL they choose to use (that is, reuse). Detecting X or Z values on block boundaries can significantly reduce debug during system-level integration of multiple blocks or IP. The *X detection* pattern is useful for detecting X propagation as well as unconnected ports.

Solution. Detect unconnected bits, undriven bits, and X propagation in Verilog as follows:

```
^<expression> === 1'bX
```

For example, using the OVL assert_never monitor, you can dctect any bit for the Verilog expression `expr[3:0]` that is unconnected or undriven as follows:

Example 6-1 *OVL* X detection

```
assert_never invalid (clk, reset_n, ^expr[3:0] === 1'bX);
```

Figure 6-1 demonstrates that `expr[3]` is unknown after an active low reset.

Figure 6-1 expr[3] is undriven or unconnected

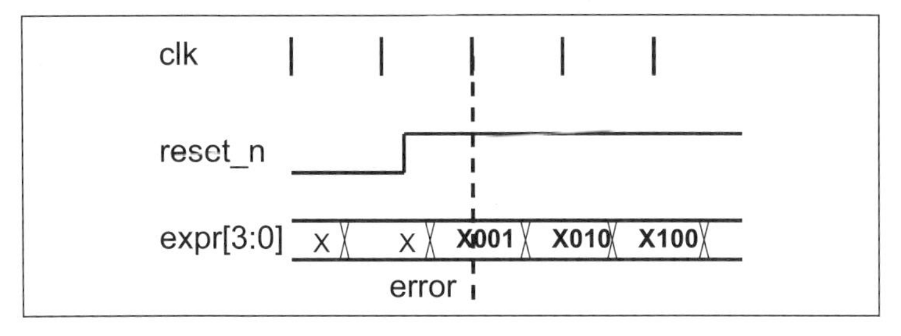

The OVL assertion in Example 6-1 quickly detects, isolates, and reports this problem. For instance, if the violation occurs at time 150 and is detected in the hierarchy `top.my_mod` (with the assert_never OVL example shown in Example 6-1), then the following default message prints during simulation:

1. Lint tools are superior to assertions for detecting unconnected signals and, in general, are the authors' recommended method for identifying these problems.

OVL_FATAL : ASSERT_NEVER : VIOLATION : : severity 0 : time 150 : top.my_mod.invalid

Note that you can override the default error message by creating a unique (that is, customized) error message for each instantiated OVL assertion module. The customized error message is passed into the instantiated OVL as a string through the *msg* parameter as shown in Example 6-2. Note, the first parameter in this example is the *severity level*, while the second parameter is an *options* parameter. For additional details on the OVL parameter values, see Appendix A.

Example 6-2 *OVL* X detection with customized message

```
assert_never #(1,0,"X detected")
  invalid (clk, reset_n, ^expr[3:0] === 1'bX);
```

The customized message in Example 6-2 is:

OVL_ERROR : ASSERT_NEVER : `X detected` : : severity 1: time 150 : top.my_mod.invalid

Refer to Appendix A for additional ways to customize OVL messages.

Methodology guideline

Notice in this example that the *severity level* 1 changed the severity to ERROR. We recommend that only assertions associated with the testbench have a *severity level* of 0 (FATAL), which causes simulation to halt.

SystemVerilog **$isunknown** system task

Alternatively, detecting unconnected ports or signals that were assigned an X value is accomplished in SystemVerilog using the newly defined system task $isunknown(<*expression*>), which returns true if any bit of the *expression* is X. This is equivalent to:

```
^<expression> === 1'bX
```

Example 6-3 demonstrates how to code the same OVL assertion shown in Example 6-1 using the SystemVerilog assert construct and the $isunknown system task.

Example 6-3 *SystemVerilog* undriven signal detection

```
always @(posedge clk) begin
  if (reset_n)
    invalid: assert property (!$isunknown(expr));
end
```

Considerations. Instead of specifying an assertion for each input signal, you should group an appropriate set of signals and check them as a single assertion, as shown in Example 6-4

Example 6-4 *SystemVerilog* unknown check for multiple signals

```
`ifdef X_DETECTION

always @(posedge clk) begin
  if (reset_n)
    x_prob: assert(!$isunknown({req,tras_start,addr,burst,we})) else
      $error("undriven cpu input signal req=%h trans_start=%h addr=%hburst=%h we=%h",
                    req, tras_start, addr, burst, we);
end

`endif
```

Methodology guideline

For improved simulation performance, the engineer might consider bracketing all *X detection* checks with a Verilog `` `ifdef `` compiler directive. After the model has reached a stable point during verification (that is, X values are no longer propagating), disable these assertions.

6.2.2 Valid range pattern

Pattern name. Valid range

Problem. Ensure that a multi-bit signal or expression evaluates to a value within a valid min/max range.

Motivation. Signals (and expressions) within the RTL model may incorrectly evaluate to values that are not supported within the structure of the model. For example, consider a simple FIFO with a maximum depth of six elements. If we use a three-bit pointer to track the current number of valid elements contained within the FIFO, then the pointer should never evaluate to seven. In general, range checks are specific (and critical) to a given RTL implementation.

Context. Signals within many RTL control structures, such as counters, memory address circuits, and finite-state machines (FSM), are often limited to a specified valid range. In addition, datapath circuits often require variables or expressions to evaluate within the allowed range.

Solution. You can detect multi-bit variables or expressions outside of a valid minimum or maximum range by writing the following Verilog expression:

```
expr >= min_val && expr <= max_val
```

PSL range check

For instance, in Example 6-5, we have written a PSL assertion to validate that a three-bit `fifo_depth` variable evaluates to a value between a range of zero and six.[2]

Example 6-5 ***PSL* valid range check**

```
assert always (reset_n ? (fifo_depth < 7) : 1'b1) @(posedge clk);
```

Notice the Verilog conditional expression in Example 6-5, which is used to prevent a false evaluation of the assertion during reset. In the next section, we demonstrate a simpler way to code assertions containing conditional expressions using the PSL implication operator.

Figure 6-2 illustrates a range violation that the PSL assertion in Example 6-5 would identify.

Figure 6-2 **fifo_depth variable out of valid range zero thru six**

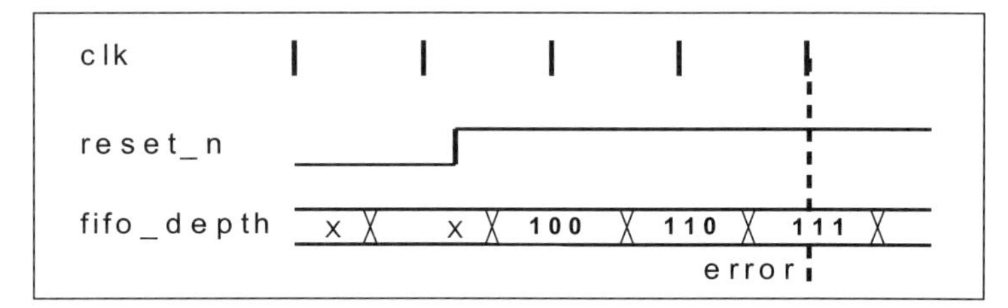

OVL range check

The OVL assert_range monitor detects range violations. In Example 6-6, the monitor reports an error if the three-bit `fifo_depth` variable evaluates outside the *min*/*max* range. In this case, the *min* and *max* values (which are parameters in the OVL module) are zero and six, respectively.

Example 6-6 ***OVL* valid range check**

```
assert_range #(0,3,0,6) above_full (clk, reset_n, fifo_depth);
```

The assert_range assertion continuously monitors the *test_expr* at every positive edge of the triggering event or clock *clk*. It contends that a specified *test_expr* will always have a value within a legal *min*/*max* range; otherwise, an assertion will fire (that is, an error condition will be detected in the code). The *test_expr* can be any valid Verilog or VHDL expression (depending on the library you are using). The *min* and *max* should be a valid parameter and *min* must be less than or equal to *max*.

2. We do not check for the case of `fifo_depth` less than zero, since all Verilog unsigned registers are greater than or equal to zero.

OVL Verilog Syntax

assert_range [#(severity_level, width, min, max, options, msg)] inst_name (clk, reset_n, test_expr);

severity_level	Severity of the failure with default value of 0.
width	Width of the monitored expression *test_expr*.
min	Minimum value allowed for range check. Default to 0.
max	Maximum value allowed for range check. Default to (2***width* - 1).
options	Vendor options.
msg	Error message that will be printed if the assertion fires.
inst_name	Instance name of assertion monitor.
clk	Triggering or clocking event that monitors the assertion.
reset_n	Signal indicating completed initialization (for example, a local copy of *reset_n* of a global reference to *reset_n*).
test_expr	Expression being verified at the positive edge of *clk*.

For additional details of the OVL **assert_range**, see Appendix A.

SystemVerilog range check

Alternatively, the range check previously demonstrated using the OVL in Example 6-6 for a three-bit `fifo_depth` variable can be expressed in SystemVerilog using the assert construct as shown in Example 6-7.

Example 6-7 ***SystemVerilog*** **valid range check**

```
// procedural assertion

always @(posedge clk) begin
  if (reset_n)
    full: assert property (fifo_depth < 7) else
             $error ("fifo_com Fifo64 Internal Failure, send mail to support@fifo.com.");
end
```

Considerations. Valid range patterns are useful in any design where a physical address limit has been established on some fully addressable space.

6.2.3 One-hot pattern

Pattern name. One-hot

Problem. Ensure that no more than one bit of a multi-bit variable or expression is active high at a time (that is, all other bits are active low).

Motivation. Often signals in RTL designs have a specific or required encoding. For example, one-hot encodings are common in high-speed designs. A common use of one-hot encoding is associated with multiplexers, as shown in Example 6-8.

Note that if a condition occurs where multiple select bits are active high in Example 6-8, then the RTL model during simulation will not reflect the actual circuit behavior (that is, the first match within the casez statement will take effect). In fact, for some vendor ASIC multiplexer cells (such as a pass-through mux), the circuit can be damaged (that is, burn up) if more than one select line is active at a time.

Context. The one-hot pattern is most useful in control circuits. It ensures that the state variable of a finite state machine (FSM) implemented with one-hot encoding will maintain proper behavior (that is, exactly one bit is asserted high). In datapath circuits, one-hot checks ensure that the enabling signals of bus-based designs do not generate bus contention.

Example 6-8 one-hot multiplexer

```
module dmux4(o, sel, i0, i1, i2, i3);
  parameter           WIDTH = 1;
  input [WIDTH-1:0]   i0, i1, i2, i3; // input data
  input [3:0]         sel;            // select signal
  output [WIDTH-1:0]  o;              // output

  always @(i0 or i1 or i2 or i3 or sel) begin
    casez (1'b1) // synopsys parallel_case
      sel[0]: o = i0;
      sel[1]: o = i1;
      sel[2]: o = i2;
      sel[3]: o = i3;
      default: $display ("No active select line on dmux4 %m");
    endcase
  end
endmodule
```

Solution. You can check for a *zero or one-hot* condition on a multi-bit variable *expr* by writing the following Verilog expression:

```
(expr & (expr - 1)) == 1'b0
```

Hence, detecting a pure one-hot condition can be expressed as:

```
(expr != 0) && ((expr & (expr - 1)) == 1'b0)
```

OVL one-hot check

Using the OVL assert_one_hot monitor, the four-bit select line `sel` in Example 6-8 can be validated as shown in Example 6-9.

Example 6-9 *OVL* one-hot check

```
assert_one_hot #(0,4) dmux4_one_hot ('TOP.clk, 'TOP.reset_n, sel);
```

Methodology guideline

Notice the multiplexer circuit shown in Example 6-8 is unclocked. To prevent false firings due to the transient behavior of events in simulation, we have chosen to use a global clock defined in the top module of the design to sample the select signal (that is, `'TOP.clk`, where the `TOP` Verilog macro can be redefined to any appropriate top module in the heirarchy). Some designs define a special assertion clock and reset signal in the top module, which can be tuned to prevent false firings of assertions during simulation. This works well for simulation, but can create problems when synthesizing assertions into actual hardware checkers [Nacif et al. 2003] or formal tools.

Figure 6-3 demonstrates the one-hot violation that the OVL assertion in Example 6-9 would identify.

Figure 6-3 dmux4 's' variable select one-hot violation

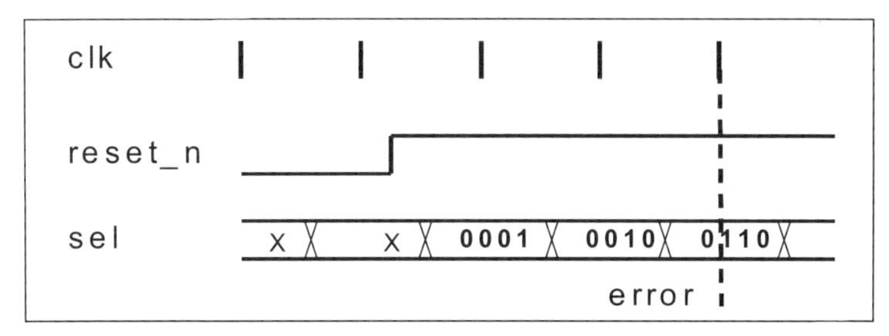

Example 6-10 *SystemVerilog* one-hot multiplexer check

```
module dmux4(o, sel, i0, i1, i2, i3);
  parameter           WIDTH = 1;
  input [WIDTH-1:0]   i0, i1, i2, i3; // input data
  input [3:0]         sel;            // select signal
  output [WIDTH-1:0]  o;              // output

  always @(i0 or i1 or i2 or i3 or sel) begin

    // procedural clocked assertion
    if ('TOP.reset_n)
      assert property (@(posedge 'TOP.clk)($countones(sel)==1))
        else $error("dmux4 select line one-hot violation at %m");

    casez (1'b1) // synopsys parallel_case
      sel[0]: o = i0;
      sel[1]: o = i1;
      sel[2]: o = i2;
      sel[3]: o = i3;
    endcase
  end
endmodule
```

SystemVerilog $countones system task

Alternatively, we could code this assertion in SystemVerilog as shown in Example 6-10. This example uses the new $countones system task, which returns an integer count for the number of bits set to one.

The $countones system task evaluates a bit-vector expression as its input argument and returns an integer value that represents the total number of bits that were identified as a one in the expression.

Considerations. In addition to validating one-hot signals, such as a mux control set of signals or one-hot encoding of state machines, the *one-hot pattern* is useful for validating mutually exclusive events. For instance, assume we have a design that contains three separate memory controller finite state machines (FSMs). If only a single memory controller is permitted to be in a write mode state at a time, we could check this condition using the OVL assert_zero_one_hot monitor as follows:

Example 6-11 *OVL* assert_zero_one_hot 3-bit mutual exclusive event check

```
assert_zero_one_hot #(0,3) mutex_wr_state (clk, reset_n,
      {FSM_1=='WR_STATE, FSM_2=='WR_STATE, FSM_3=='WR_STATE});
```

Example 6-11 parameterizes assert_zero_one_hot monitor to a width of three bits to check the mutual exclusivity of the three FSMs. Notice that a Verilog concatenation expression is created, with each bit representing one of the mutually exclusive events (that is, a current write mode status for each of the three FSMs).

6.2.4 Gray-code pattern

ensure proper gray-code encoding for queue pointers

Pattern name. Gray-code

Problem. Ensure that only a single bit changes value between clock transitions.

Motivation. To ensure data integrity when transferring queue pointers between different clock domains, often the pointer value is encoded with a gray-code. Hence, only a single bit of the gray-code encoded pointer is permitted to change per clock transition. This encoding helps identify possible skew issues between the multiple bits during an asynchronous transfer.

Context. There are a number of problems to be addressed when passing information across clocking domains. These are: metastability, fast to slow transfer (and vice versa), and timing skew between multiple bits. Two of these problems are solved by use of proper synchronization logic. The third problem can be minimized (and identified) through gray-code encoding of the transferred multi-bit value.

Solution. To ensure proper gray-code transitions, we can specify an assertion as shown in Example 6-12.

Example 6-12 ***SystemVerilog* assertion for gray-code encoding**

```
property legal_graycode(code);
  @(posedge clk) ($countones($past(code) ^ code)<=1));
endproperty
assert legal_graycode(async_ptr);
```

Note that the exclusive OR function computes the changing bits between the previous and current value of code.

6.3 Set patterns

In this section, we define assertion patterns that evaluate a group of signals or a bit-vector expression. The evaluation (that is, value) of this group of signals must be contained within a set of valid possible choices. Examples of *set patterns* include valid bus tags, valid opcodes, and any valid encoding defined within an RTL model.

6.3.1 Valid opcode pattern

Pattern name. Valid opcode

Problem. Ensure that an RTL bit-vector signal (expression) evaluates to a value that is contained within a set of possible legal values.

An ALU opcode must evaluate to a valid values

Motivation. In RTL design, multiple signals are often grouped by type, which permits information encoding. If the signal group evaluates to a value outside the set of legal values, then an error can occur within the design. Consider an opcode sent to an ALU. The ALU decodes the opcode to control its operations. If an invalid opcode value is decoded, it could result in an error condition or unpredictable behavior in the design (that is, if the ALU hardware doesn't trap the illegal case).

Context. The *valid opcode pattern* is useful in control or other circuits that contain a finite set of encoded commands or opcodes.

Solution. Ensure that an opcode evaluates to a valid set of values by writing an explicit expression to check each legal value. For example, assume a three-bit opcode was encoded with the following commands:

```
'define ADD=1
'define SUB=2
'define RD=3
'define WR=7
```

Hence, the values in the set (0, 4, 5, 6) are not valid. For this simple case, write a Verilog expression to validate the opcode values as follows:

```
(opcode=='ADD) || (opcode=='RD) ||
(opcode=='SUB) || (opcode=='WR)
```

OVL set check

Using the OVL assert_always monitor, you can validate the opcode encoding for this simple case as shown in Example 6-13:

Example 6-13 *OVL* valid opcode check

```
assert_always valid_opcode (clk, reset_n,
  (opcode=='ADD) || (opcode=='RD) || (opcode=='SUB) || (opcode=='WR));
```

SystemVerilog $inset system task

Alternatively, check that a signal evaluates to a value contained within a set with the newly defined SystemVerilog $inset system task. This system task takes in a Verilog *expression* as its first argument and a comma-separated list of *constants* as additional arguments. The task compares the value of the expression with the list of constants. If the expression evaluates to one of the constraint arguments, then a 1 is returned. However, if the expression does not evaluate to one of the constraint arguments, then a 0 is returned. Example 6-14 demonstrates a SystemVerilog assertion you can use to check for valid values of an opcode:

Example 6-14 *SystemVerilog* valid opcode check with $inset

```
// concurrent assertion

valid_op: assert property (@(posedge clk) disable iff (reset_n)
               $inset(opcode, 'ADD,'RD,'SUB,'WR)) else
              $error("CTL sent illegal opcode (%0h) to ALU.", opcode);
```

Considerations. While the *valid opcode* pattern is useful for simple opcodes, there are times when a larger set of signals can produce an incongruous value, which would be hard to enumerate using the solution recommended by this pattern. The *valid signal combination* pattern demonstrates a technique for validating that a group of signals evaluates to a valid set of values.

6.3.2 Valid signal combination pattern

Pattern name. Valid signal combination

Problem. Ensure that a combination of signals evaluates to only legal (acceptable) values defined within a set.

Motivation. RTL designers frequently group multiple signals to convey a specific piece of information. In other words, the relationship between the set of signals must be consistent to convey a proper meaning. Incorrect combinations of signals can lead to inaccurate and unintended operations by the receiver. Consider a processor-to-memory interface with signals that describe `read`, `write`, `burst` (that is, cache line operation), `size` (that is, byte, halfword, word), and *write through* `wt` (that is, writing directly through to memory). For this set of signals, specific combinations are illegal, or only a few combinations may be legal.

Context. The *valid signal combination* pattern is useful when the combination of individual signals convey a particular meaning, which are then assembled and sent to another unit. Examples include bus interfaces, control interfaces, and status buses.

Example 6-15 *OVL* valid signal combination check

```
`ifdef ASSERT_ON

reg trans_ok;

always @(read or write or burst or size or wt)
  casez({read, write, burst, size, wt})
    6'b1_0_1_00_?, // cache (burst) read.
    6'b1_0_0_00_?, // single byte read
    6'b1_0_0_01_?, // halfword read
    6'b1_0_0_11_?, // word read
    6'b0_1_1_00_0, // cache (burst) write
    6'b0_1_0_00_0, // single byte write
    6'b0_1_0_01_0, // halfword write
    6'b0_1_0_11_0, // word write
    6'b0_1_0_00_0, // single byte writethru
    6'b0_1_0_01_0, // halfword writethru
    6'b0_1_0_11_0, // word writethru
    6'b0_0_0_00_0: // nothing.
                   trans_ok = 1'b1;
    default:       trans_ok = 1'b0;
  endcase

  // OVL assertion
  assert_always illegal_mem_req (clk, reset_n, trans_ok);

`ifdef ASSERT_OFF
```

Solution. To specify the valid values associated with the grouping of a set of signals, you should create a table for specifying the legal combinations of signal values. This table can be expressed using a Verilog case/casez statement or the SystemVerilog $inset or $insetz system task. The case/casez statement assigns a variable, which is then checked by an OVL monitor to identify any illegal combinations, as shown in Example 6-15.

Example 6-16 shows a SystemVerilog concurrent assertion that is used to check a processor-to-memory interface for a valid combination of signal values.

Note that on some lines, the casez *case_item* alternative contains a don't care matching character (? representing Z), which enables us to express the legal combinations in a compact form. This makes the tables regular and improves readability. In Example 6-16, the SystemVerilog $insetz system task enables us to achieve the same type of check as the casez structure demonstrated in Example 6-15.

Figure 6-4 illustrates the usefulness of checking illegal signal combinations on a CPU bus. For this example, a cache `read burst` precedes a cache `write burst`, which precedes an illegal active `read` and `write` signal. To reduce debug time, it is best to isolate illegal combinations close to the source of the error, as opposed to depending on the effect of the illegal combination to propagate to an observable point (for example, an output port).

Example 6-16 ***SystemVerilog* legal signal combination check**

```
//declarative assertion

assert property ( @(posedge clk)
  $insetz({read, write, burst, size, wt},
          6'b1_0_1_00_?,    // cache (burst) read.
          6'b1_0_0_00_?,    // single byte read
          6'b1_0_0_01_?,    // halfword read
          6'b1_0_0_11_?,    // word read
          6'b0_1_1_00_0,    // cache (burst) write
          6'b0_1_0_00_0,    // single byte write
          6'b0_1_0_01_0,    // halfword write
          6'b0_1_0_11_0,    // word write
          6'b0_1_0_00_0,    // single byte writethru
          6'b0_1_0_01_0,    // halfword writethru
          6'b0_1_0_11_0,    // word writethru
          6'b0_0_0_00_0))   // nothing.
  else $error("Illegal memory request {read,write,burst,size,wt}=%0h",
                {read, write, burst, size, wt});
```

Figure 6-4 **Illegal read/write burst**

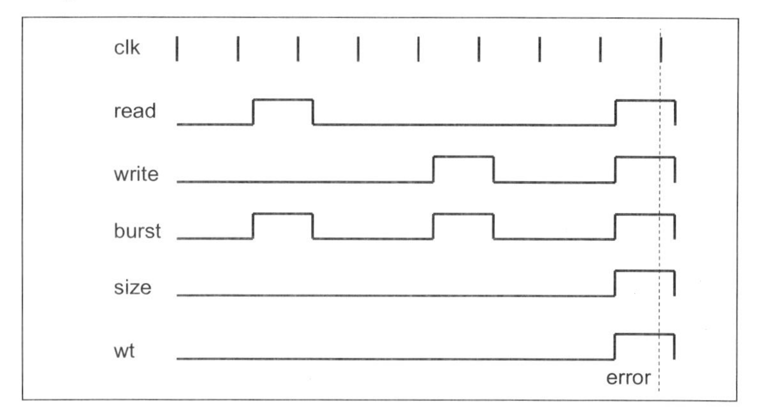

Considerations. This pattern is useful in many situations in your design to validate what you consider a legal state, illegal states, or valid input requests. The combination signals may be from dissimilar sources, which combine to produce an incoherent request. Sometimes, the legal combinations outnumber the illegal combinations, which means that an evaluation contained within the illegal might be simpler to express, as demonstrated by the *invalid signal combination* pattern.

6.3.3 Invalid signal combination pattern

Pattern name. Invalid signal combination

Problem. Ensure that a combination of signal values do not evaluate to values specified within a set of values.

Motivation. It is common to see combinations of two or more signals that should never be active at the same time. It is also common to have a grouping or set of signals, which should not evaluate to a combination of values. Defining these illegal relationships not only aids other readers by documenting expected behavior, it also reminds the designer of important characteristics of the design that must be considered (and not ignored) during future logic optimization or bug fixes.

Context. The *invalid signal combination* pattern is useful when applied to a grouping of individual signals, which are then assembled and sent on to another unit to convey a particular meaning or control. Examples include bus interfaces, control interfaces, and status buses.

Example 6-17 ***SystemVerilog*** **illegal signal combination check**

```
// declarative assertion

assert property ( @(posedge clk) disable iff (reset_n)
   not $insetz({read, write, burst, size[1:0], wt},
       6'b1_0_1_01_?, 6'b1_0_1_10_?, // wrong size
       6'b1_0_1_11_?,
       6'b0_1_1_01_0, 6'b0_1_1_10_0, // wrong size
       6'b0_1_1_11_0,
       6'b0_1_1_00_1, 6'b0_1_1_01_1, // burst writethru
       6'b0_1_1_10_1, 6'b0_1_1_11_1,
       6'b1_0_0_10_?, 6'b0_1_0_10_0))// wrong size.
  else $error("Illegal memory request {read,write,burst,size,wt}=%0h",
               {read, write, burst, size, wt});
```

Solution. Check for illegal signal value combinations by writing an expression that represents the combinations that should not occur. A table approach (like the previous *valid signal combination* pattern) is useful for larger sets of signals. Example 6-15 shows a valid CPU bus check that has been modified in Example 6-17 to check for the illegal signal value combinations.

Note that in this example the number of elements in the illegal set is the same as that expressed in the legal set. However, in many cases the illegal set is easier to express than the valid set.

For another simpler case of the *illegal signal set* pattern, consider an assertion for a design that contains a cache, as shown in Example 6-18. This cache has a read, write, invalidate, and a flush port. For this particular design, it is illegal for an active invalidate and a flush signal to occur at the same time:

Example 6-18 ***SystemVerilog*** **illegal cache invalidate and flush request**

```
// declarative assertion

assert property ( @(posedge clk) disable iff (reset_n)
    not (invalidate & flush))
  else $error("Cache received illegal invalidate and flush request.");
```

This design may also forbid writing during invalidation or flushing (or both). The equation is easily extended to include more cases of illegal behavior.

Considerations. Additional illegal patterns may occur when considering combination of signals spanning across multiple cycles (for example, a previous cycle or a future cycle). See section 6.5 on page 185 to detect these types of illegal operations.

6.4 Conditional patterns

Conditioned logic is common in RTL design. For example, signals with names like `reset`, `enable`, `valid`, `ready`, `request`, `ack` and `done` often trigger conditions for process activation. In this section we demonstrate *conditional expression* patterns that utilize a controlling signal (or signals) to define precisely when a specific requirement must be checked. In addition, we demonstrate *sequence implication* patterns where the conditional event is a sequence of Boolean expression.

6.4.1 Conditional expression pattern

Pattern name. Conditional expression

Problem. Many designs use a `valid` or `enablc` signal to indicate the proper time when information is available for processing. Hence, validating the correctness of the received data is dependant on the status of the conditional expression.

Motivation. Returning to the examples in Section 6.3.1 "Valid opcode pattern", we extend the valid opcode pattern to use a signal *valid* to indicate when the opcode is to be use to analyze the data coming to an ALU block.

Context. Apply the *conditional expression* pattern to any design containing enabled or conditional logic.

Solution. Check enabled or conditional logic by using the new SystemVerilog implication operator as shown in Example 6-19. For this example, legal opcodes are checked only when the *enable* signal is active.

Example 6-19 ***SystemVerilog*** **conditional checking of valid opcode**

```
// declarative assertion

assert property ( @(posedge clk) disable iff (reset_n)
    (enable |-> $inset(opcode, 1, 2, 3, 7)))
  else $error("CTL sent illegal opcode (%0h) to ALU.", opcode);
```

In math, the implication operator consist of an *antecedent* that implies a *consequence* (for example, `A -> C`, which reads `A` implies `C`). If the antecedent is true, then the consequence must be true for the implication to pass. If the antecedent is false, then the implication passes regardless of the value of the consequence. Similarly, the SystemVerilog implication operator `|->` allows you to state a Boolean expression or a prerequisite sequence as an antecedent. When the Boolean expression or prerequisite sequence is satisfied, then this implies that the consequence Boolean expression or suffix sequence must be satisfied.

Figure 6-5 **Illegal opcode 5 to ALU**

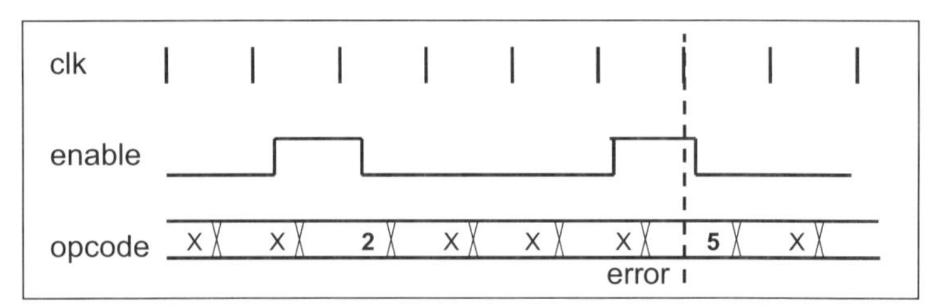

Figure 6-5 demonstrates an error case when an illegal opcode of 5 occurs at the same time an `enable` signal is active. This case could be caught with the *conditional expression* pattern assertion described in Example 6-19.

As another example, consider the case for the four-bit register (`reg_select`) that is expected to evaluate to a one-hot value whenever the `valid_read` signal is active, as shown in Example 6-20.

Example 6-20 ***SystemVerilog* legal selection check during valid operation**

```
// declarative assertion

assert property ( @(posedge clk) disable iff (reset_n)
    (valid_read |-> $countones(reg_select)==1))
  else $error("REG block performed illegal register selection (%b)", select);
```

As another example of a *conditional expression* pattern, we re-code Example 6-15 to check for a valid transaction on a legal combination of signals using the implication operation, as shown in Example 6-21.

Finally, the X detection patterns presented in Section 6.2.1 might be coded with an implication if the check is dependant on a conditional expression as shown in Example 6-22.

Considerations. These examples demonstrate how to use the interaction of conditional expression within your design to validate inputs, outputs, and internal states. However, for some assertions, combinatorial interactions may be insufficient to specify the enabling condition. In Section 6.4.2 "Sequence implication pattern", we discuss assertions that consider sequence triggering conditions for implication.

Example 6-21 *SystemVerilog* check for valid transaction

```
// declarative assertion

assert property ( @(posedge clk) disable iff (reset_n)
  (trans_start |-> $insetz({read,write,burst,size,wt},
          6'b1_0_1_00_?,    // cache (burst) read.
          6'b1_0_0_00_?,    // single byte read
          6'b1_0_0_01_?,    // halfword read
          6'b1_0_0_11_?,    // word read
          6'b0_1_1_00_0,    // cache (burst) write
          6'b0_1_0_00_0,    // single byte write
          6'b0_1_0_01_0,    // halfword write
          6'b0_1_0_11_0,    // word write
          6'b0_1_0_00_0,    // single byte writethru
          6'b0_1_0_01_0,    // halfword writethru
          6'b0_1_0_11_0,    // word writethru
          6'b0_0_0_00_0))) // nothing.
  else $error("Illegal request {read,write,burst,size,wt}=%0h",
          {read, write, burst, size, wt});
```

Example 6-22 *SystemVerilog* check for undriven data when valid

```
// declarative assertion

assert property ( @(posedge clk) disable iff (reset_n)
    (data_valid |-> !$isunknown(data[31:0]))
  else $error("Undriven data bus (%h) during data return",
          data[31:0]);
```

6.4.2 Sequence implication pattern

Pattern name. Sequence implication

Problem. A combinational condition within a design (for example, a simple active `enable` or `valid` signal) may be insufficient to describe the triggering event required for a conditional pattern. A *prerequisite sequence* might be required as a triggering event for an assertion check.

Motivation. Bus protocols generally contain some kind of arbitration scheme where an active `grant` is generated (giving permission to use the bus) after an active `request`. Consider a system where a bus client generates a `request` signal, then eventually receives a `grant` signal, and finally is expected to activate a `start` signal to begin a transaction in the cycle immediately after the `grant`. In this system, the prerequisite sequence condition *"a request followed by a grant"* implies an active `start` signal in the next cycle. The *sequence implication* pattern addresses this type of system.

Context. Apply the sequence implication pattern to complex protocols between a set of blocks, where a block's specified input sequence (that is, a prerequisite sequence) generates an expected result (that is, a suffix sequence). The *sequence implication* pattern may also be applied to inter-block communications where a sequence of internal states triggers a check for an expected result.

Solution. A complex event, such as a prerequisite sequence, can be used to define a triggering event for a design. When the triggering event occurs, then some other condition within the design must be valid. For example, we can write an assertion that specifies the following *sequence implication* pattern:

> Whenever an active *grant* is received—then on the following cycle, an active *start* must occur—provided that the initial *request* was never removed. In addition, the *grant* must be received within three cycles after the initial request.

Figure 6-6 illustrates the legal case where a bus request was made, as shown by the active `req` signal. The active `grant` occurs on the third cycle after the initial `req`. However, the bus request was removed prior to the start of the transaction, which means that an active `start` signal must not occur on the cycle immediately after the `grant`.

Figure 6-6 **Legal sequence implication pattern**

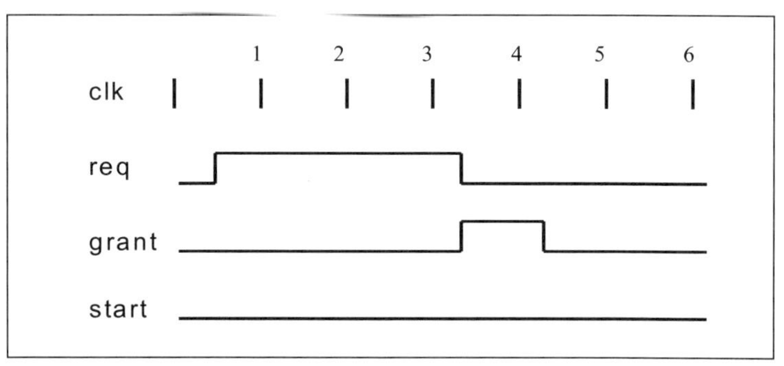

Figure 6-7 illustrates an illegal case where a bus transaction request was made as shown by the active `req` signal. The active `grant` is generated on the third cycle after the initial `req`. (that is, clock tick 4). However, since the initial `req` is still active, an error occurs at clock tick 5 since an active `start` did not occur on the immediate cycle after an active `grant`.

Figure 6-7 **Illegal sequence implication pattern**

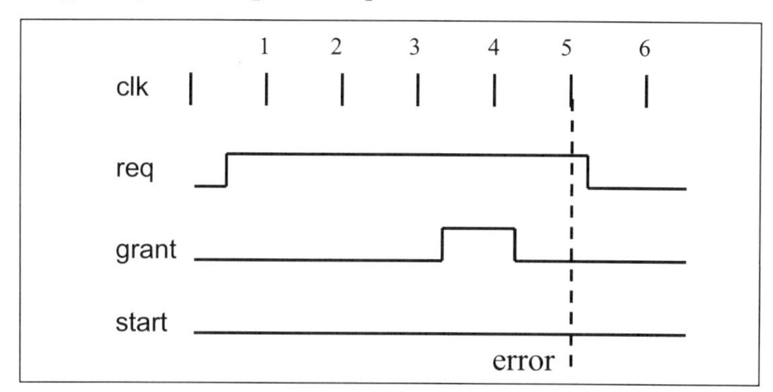

Example 6-23 demonstrates how to write a PSL assertion to specify the legal behavior described in the previous example.

Example 6-23 ***PSL*** `grant` **to transaction request assertion**

```
// declarative assertion

// A start can only occur after a grant for an active request

assert always {{req[*1:4]}:{grant}; req} |-> {start} @(posedge clk);
```

For this example, we assert that if a prerequisite sequence (involving the `req` and `grant` signals) is satisfied, then an active `start` signal is required. To specify a time limit where an active `grant` occurs within four cycles after an initial `req`, we use the PSL sequence *fusion operator* (:). The fusion operator enables us to describe a sequence in which the ending event for the operator's left-hand sequence overlaps with the starting event for its right-hand sequence. Hence, for our example, we are specifying that an active `grant` overlaps with the last active `req` in a set of sequences of one, two, three, or four consecutive `req`s. Figure 6-8 demonstrates an example where a `grant` overlaps the final cycle in a sequence consisting of four consecutive `req`s, at clock tick 4.

Figure 6-8 **Ending event in sequence overlaps with consequence**

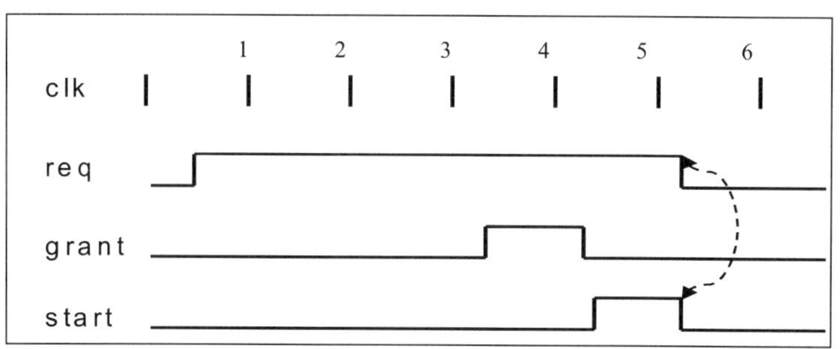

Note that Example 6-23 uses the PSL *weak suffix implication operator* (|->), which does not advance time between the *prerequisite* and *suffix sequences* (that is, there is no new clock tick). This enables us to specify that the ending cycle in the prerequisite sequence (that is, the final active `req` signal) must overlap with the starting cycle of the suffix sequence (that is, an active `start` signal). This situation is demonstrated in Figure 6-8 at clock tick 5.

See Appendix B for additional details on the PSL *weak suffix implication operator*.

Note that for the previous example, to assert that `grant` must occur within four cycles of an initial `req` (regardless of whether the `req` is removed), we could write:

Example 6-24 *PSL* `grant` timeout check

```
// declarative assertion

// assert grant is recieved within 3 clocks after an initial req

assert always {rose(req)} |-> {{[*1:4]}:{grant}) @(posedge clk);
```

The PSL sequence fusion operator in Example 6-24 enables us to specify a *set* of suffix sequences where a `grant` occurs on the first, second, third, or fourth cycle of one of the sequences.

For example, the fusion of the sequences `{{1[*1:4]}:{grant}}` cxpands into thc following set of sequences:

```
(a)          {grant}
(b)          {1;grant}
(c)          {1;1;grant}
(d)          {1;1;1;grant}
```

Where the "1" represents *true*, which allows us to match a signal or Boolean expression and advance time to the next cycle. For case (a), where the `grant` occurs on the first cycle of the suffix sequence, the `grant` will actually overlap the initial `req` of the prerequisite sequence shown in Example 6-24. For case (d), where the `grant` occurs on the fourth cycle of the suffix sequence, the actual `grant` occurs on the third cycle immediately after the initial `req`.

Considerations. To apply sequence implication patterns to your design using the OVL, consider using the assert_next and assert_cycle_sequence monitors.

In addition to specifying bus protocol assertions, the *sequence implication* pattern is useful for specifying various control logic assertions where a prerequisite sequence is represented as a valid progression of valid states and input values.

6.5 Past and future event patterns

Generally, a previous event within a system places requirements on the system's current state. Similarly, the current state in a system places a requirement on future events. The choice of specifying the relationship between a past or future event to a current state in the system is a matter of convenience. In this section, we introduce *past event* and *future event* patterns to describe these relationships.

6.5.1 Past event pattern

Pattern name. Past event

Problem. Detecting incorrect behavior for the current state of a system often depends on a previous event within the system.

Motivation. Cache protocols often have an *invalidate* command, which is used to mark the cache data as invalid for use (that is, due to a memory update somewhere in the system, the local cache data is no longer valid). If a cache read (that is, *hit*) occurs after an *invalidate*, then an error occurs (that is, there was an attempt to read invalid data). The *past event* pattern is useful for identifying invalid protocol errors.

Context. When designing an FSM, control circuit, or bus interface, a previous event of the system often influences the current state. These relationships must be validated.

Solution. Detect incorrect dependencies between a past event and the current state by referencing a previous combination of signal values within the design. SystemVerilog provides a means to access a previous combination of signal values within the verification environment through the use of the $past system function:

$past (*bit_vector_expr* [, *number_of_ticks*])

The *number_of_ticks* argument specifies the number clock ticks used to retrieve the previous value of *bit_vector_expr*. If *number_of_ticks* is not specified, then it defaults to one.

Example 6-25 demonstrates a *past event* pattern for the case where a cache hit must never occur when an invalidate occurred in the previous cycle.

Example 6-25 *SystemVerilog* past event pattern for illegal cache transaction

```
// declarative assertion

assert property ( @(posedge clk) disable iff (reset_n)
    not ($past(invalidate) & hit))
  else $error("Cache hit occurred while previous invalidate active");
```

Example 6-26 demonstrates another *past event* pattern. For this example, the bus interface starts a memory transaction by activating a request (for example, a single pulse of a req_valid signal). Once this occurs, the address bus (addr) must not change its value until the new request occurs. The PSL code in Example 6-26 demonstrates how to apply this pattern to ensure that a memory address bus is stable:

Example 6-26 *PSL* past event pattern *to* check for stable signal

```
// declarative assertion

// assert that address will not change values after a request is made.

assert always !req_valid -> prev(addr)==addr @(posedge clk);
```

Note that the assertion in Example 6-26 uses the PSL prev built-in-function to retrieve the value of addr from the previous clock cycle (defined by @(posedge clk)) in the same way as SystemVerilog $past system task demonstrated in Example 6-25 was used to retrieve the previous value of the invalidate signal.

Figure 6-9 Address changed without a request

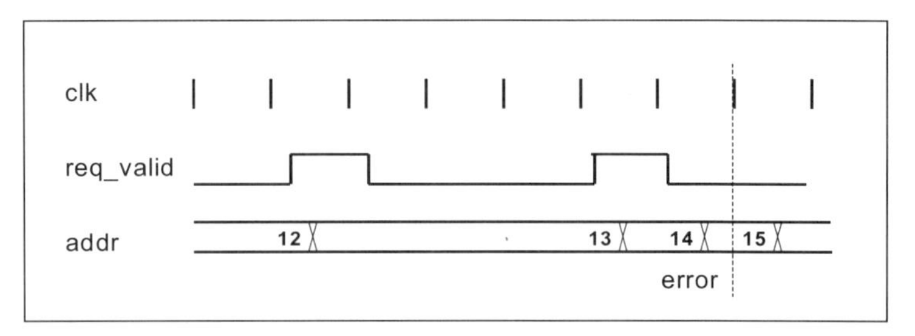

Figure 6-9 illustrates a failure for the assertion specified in Example 6-26.

Methodology guideline

Considerations. Take care when coding assertions that use either the PSL prev built-in-function or SystemVerilog $past system task to ensure that the past referenced value is valid at the current verification point in time (for example, you should not reference values prior to the start of simulation). Apply a *conditional* pattern to ensure that the past referenced time is a valid value.

6.5.2 Future event pattern

Pattern name. Future event

Problem. The current state of a system often places obligations or expectations that must be validated when some future event within the system occurs.

Motivation. Validating single-cycle and multi-cycle pulse widths on bus interfaces is critical for proper protocol behavior. For example, a protocol might place a requirement on a bus interface that a *request* signal must never be active for more than a single cycle.

Context. Bus interface protocols present many opportunities to apply the *future event* pattern. In addition, internal cycle-based timing relationships between multiple signals within the design (for example, `valid`, `flush`, and `restart`) are excellent *future event* pattern candidates.

Solution. The *future event* pattern example we present in this section is an extension of the *conditional* pattern discussed in Section 6.4. PSL, OVL, and SystemVerilog all provide a convenient means of specifying a requirement for a future event, which is dependant on the current state of the system. For example, use the PSL `next` operator to specify that the `req_valid` is always inactive during the cycle immediately after it was activated, as shown in Example 6-27.

Example 6-27 ***PSL*** **future event pattern check for a single cycle pulse**

```
// declarative assertion
// assert req_valid will never be active high for more than 1 cycle

assert always req_valid -> next !req_valid
          abort !reset_n @(posedge clk);
```

The PSL assertion in Example 6-27 can be coded with an OVL assertion as shown in Example 6-28:

Example 6-28 ***OVL* future event pattern check for a single cycle pulse**

```
// assert req_valid will never be active high for more than 1 cycle

assert_next one_req (clk, reset_n, req_valid, !req_valid);
```

Example 6-29 and Figure 6-10 are examples of a *future event* pattern related to a cache protocol. For this example, whenever an active `invalidate` occurs, then a cache `hit` must never occur on the next cycle.

Example 6-29 ***PSL* cache `invalidate/hit` check using the next operator**

```
// declarative assertion
// assert that a cache hit never occurs after an invalidate

assert never invalidate -> next hit @(posedge clk);
```

Figure 6-10 **Illegal cache `hit` after `invalidate`**

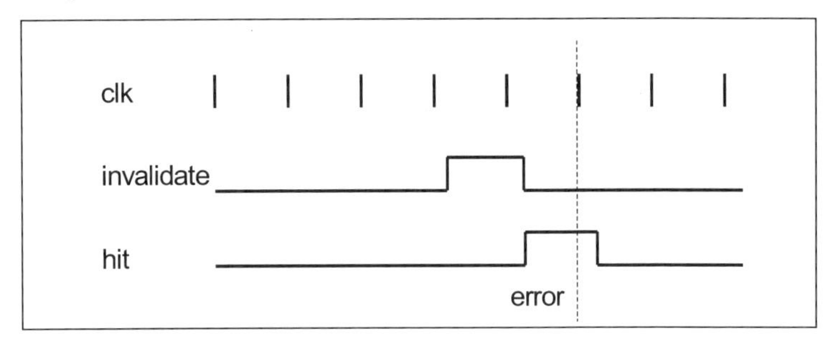

Considerations. Validating a future event is not limited to specifying events on subsequent cycles. For example, we might have a requirement: the cache must never return a `hit` within four cycles after an `invalidate`. Example 6-30 demonstrates this multi-cycle future event pattern using a PSL repetition operator for the `hit` signal (that is, [*4]).

Example 6-30 ***PSL* future event pattern check for 4 cycles of no `hit`**

```
// declarative assertion
// assert cache will not return a hit within 4 cycles after invalidate

assert always {invalidate} |=> {!hit [*4]} @(posedge clk);
```

The PSL assertion in Example 6-30 and the OVL in Example 6-31 achieve the same purpose. That is, they specify a requirement for four inactive `hit` cycles after `invalidate`.

Example 6-31 *OVL* future event pattern check for 4 cycles of no `hit`

```
// declarative assertion
// assert cache will not return a hit within 4 cycles after invalidate

assert_cycle_sequence #(0,4) inv_hit
               (clk, `TRUE, invalidate, {4{!hit}});
```

Example 6-32 also specifies a requirement for four inactive `hit` cycles after `invalidate`.

Example 6-32 *SystemVerilog* future event pattern check for 4 cycles of no `hit`

```
// declarative assertion
// assert cache will not return a hit within 4 cycles after invalidate

assert property (@(posedge clk) disable iff (reset_n)
   (invalidate |=> !hit [*4]));
```

6.6 Window patterns

In this section, we define a set of patterns related to bounded events that reciprocally affect or influence each other. We refer to a bounded requirement on a transaction as a window, since the specification window bounds are defined by an initial starting event and conclude with either a specified time limit (that is, number of cycles) or an ending event. For example, within a window of time after an initial starting event, assert that a control signal must change its value. We discuss this type of assertion in the section for *time-bounded window* patterns. Alternatively, specify a range of time or window that terminates with an ending event. We discuss this type of assertion in the section for *event-bounded window* patterns.

6.6.1 Time-bounded window patterns

Pattern name. Time-bounded window

Problem. Ensure that logic in a design reacts to a transaction within a specified number of cycles.

Motivation. For performance reasons, or to satisfy a specified protocol requirement, often logic must be designed to react to a transaction within a specified limit of time. For example, many protocols are initiated with a *request* and conclude with an *acknowledge* within a specified number of cycles. To facilitate rapid debug for these protocols, check for a maximum timeout condition on the *acknowledge* event. These assertions help isolate problems such as a stalling bus transaction caused by a deadlock.

Context. Apply the *time-bounded window* pattern in control circuits to ensure proper synchronization of events. Common usage includes:

- verify multi-cycle data operations with an enabling condition
- verify single-cycle operations with data loaded on different cycles
- verify synchronizing conditions that require stable data after a specified initial triggering event

Solution. Specify a time limit for a transaction by defining a sequence (or set of sequences) that limits the response recognition to a fixed number of cycles. For example, if a bus interface requires that an `ack` must occur within 100 cycles after a `req`, then the PSL assertion in Example 6-33 will validate that the response does not occur outside the time window.

Example 6-33 *PSL* time limit sequence check

```
// declarative assertion
// assert that ack must occur within 100 cycles after a req.

assert always req -> next {{[*1:100]}:{ack}} @(posedge clk);
```

Note that Example 6-33 uses the PSL *fusion* operator (**:**), which allows us to specify a single-cycle overlap between the ending-cycle of the left-hand sequence and the starting-cycle of the right-hand sequence. Our objective is to specify a set of `ack` sequences ranging from a length of one cycle up to a limit of 100 cycles, which would satisfy the `req` implication. The first sequence regular expression (that is, `{[*1:100]}`), is a shortcut representation of this range:

```
{1}, {1;1}, {1;1;1}, {1;1;1;1}, {1;1;1;1}, . .
```

In this sequence, "1" followed by a semicolon allows us to advance the clock tick without any obligation (that is, we do not need to satisfy any particular Boolean expression when matching sequences at that particular point in time). Hence, fusing `{[*1:100]}` with `{ack}` enables us to match the following sequences:

```
{ack}, {1,ack}, {1,1,ack}, {1,1,1,ack}, . . .
```

Figure 6-11 illustrates an error that would be detected by the PSL assertion specified in Example 6-24.

Figure 6-11 **Time limit sequence error**

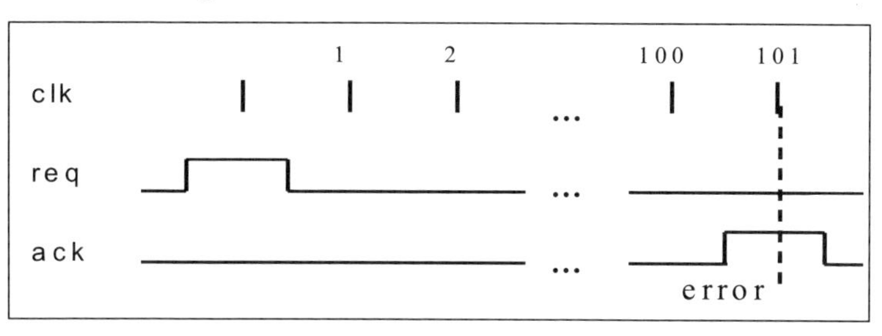

For cases when the acknowledge can occur in the same cycle as the request, as shown in Figure 6-12, the previous assertion can be re-coded, as shown in Example 6-34, with the PSL *weak suffix implication operator* (|->). This operator permits an overlap of the ending cycle of the prerequisite sequence (that is, `req`) with the first cycle in the suffix sequence. Hence, we must define a suffix sequence of length 101, since the first cycle in the sequence potentially overlaps with the `req`, followed by up to 100 additional cycles in which `ack` could occur.

Figure 6-12 **Overlapping `req` and `ack`**

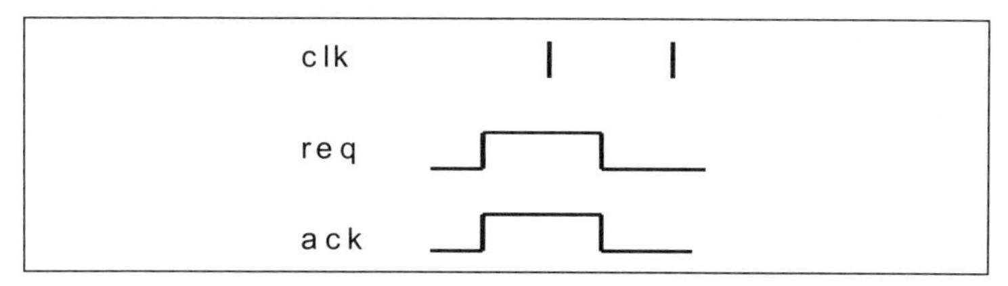

Example 6-34 ***PSL* time limit sequence check with single-cycle overlap**

```
// declarative assertion
// assert that ack must occur within 100 cycles after a req.
// the ack can overlap with the req

assert always {req} |-> {{[*0:100]}:{ack}} @(posedge clk);
```

Use the OVL assert_frame monitor in Example 6-35 to validate a timeout condition, as previously specified in Example 6-34.

The assert_frame specifies a time window (that is, a frame), which is used to limited the response. See Appendix A for additional details for the OVL assert_frame.

Example 6-35 *OVL* time limit sequence check

```
// OVL assertion
// assert that ack must occur within 100 cycles after a req.

assert_frame #(0,0,100) req_ack (clk, 1, req, ack);
```

A SystemVerilog version of this assertion is shown in Example 6-34.

Example 6-36 *SystemVerilog* time limit sequence check

```
// declarative assertion

assert property ( @(posedge clk) disable iff (reset_n)
    (req |-> ##[1:100] ack)) else
  $error ("acknowlede did not occur within 100 cycles after request");
```

Considerations. To apply time-bounded window patterns to your design using the OVL, consider using the following monitors: assert_change, assert_unchange, assert_frame, assert_width, assert_next, and assert_time.

Methodology guideline

In the previous example, we specified a window to limit completion of an acknowledge for a transaction. This timeout window may be somewhat arbitrary, but is helpful for identifying situations with unfilled requests. By parameterizing (or macro-defining) the timeout associated with your assertion, you can tune the assertion to accurately diagnose where a given request times out and investigate this region.

6.6.2 Event-bounded window patterns

Pattern name. Event-bounded window

Problem. Ensure that logic in a design behaves correctly within an arbitrary window of time, which is bounded by a specified *starting event* and *ending event*.

Motivation. Consider a simple interface protocol example between two blocks in a design, where a *single* pulse of `req` initiates a transaction, followed eventually by an active `ack`. That is, one block pulses a request signal to initiate a transaction. The other block, after receiving the request, eventually returns an acknowledge pulse to complete the transaction. In our simple example, a protocol requirement is that the first block cannot send another request until the first transaction has completed. Apply the event-bounded window pattern to this example to ensure proper behavior of the request signal.

Context. The *event-bounded window* pattern applies to protocol transactions verification or control logic involving data stability requirements, where the window of time for the transaction or data-stability is not explicitly stated, but is bounded by events within the design.

Solution. For protocols in which only a single transaction can be processed at a time, state a requirement that only a single pulse of `req` can occur prior to and including an `ack`. In other words, the `req` pulse defines an initial *starting event*, while the `ack` pulse defines the final *ending event*. These events bound an arbitrary window of time during which another active `req` signal must never occur. Example 6-37 demonstrates a PSL assertion for this *event-bounded window* pattern.

Example 6-37 ***PSL*** **event-bounded window pattern**

```
// declarative assertion
// a new request cannot start before the first one completes

assert always req -> next !req until_ ack @(posedge clk);
```

Figure 6-13 demonstrates a failing case that the PSL assertion specified in Example 6-37 would catch. The SystemVerilog version of this assertion is demonstrated in Example 6-38.

Figure 6-13 **Invalid multiple request.**

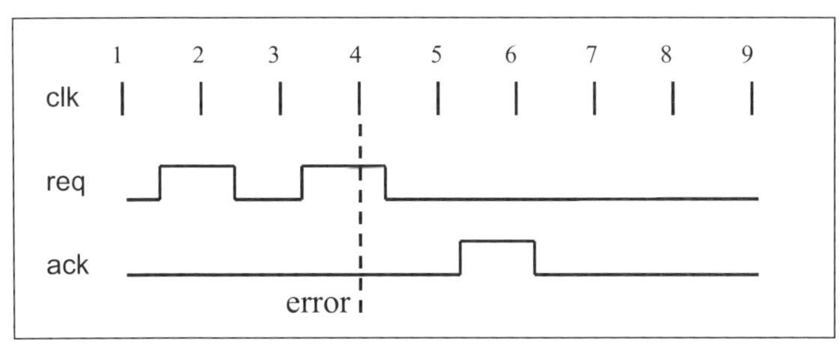

Example 6-38 ***SystemVerilog*** **event-bounded window pattern**

```
// declarative assertion
// a new request cannot start before the first one completes

assert property (@(posedge clk) disable iff (reset_n)
  (req |=> (!req throughout ack [*-> 1])));
```

In general, the SystemVerilog *nonconsecutive exact repetition (goto repetition)* operator [*-> 1] is intended to cover the semantic which matches the nth occurrence of an event (for example, [*-> 2] would match (a, true, a), see Chapter 3 and Appendix C for additional details on the SystemVerilog repetition operators).

Considerations. To apply event-bounded window patterns to your design using the OVL, consider using assert_win_change, assert_win_unchange, and assert_window. Also note that the *event-bounded window* pattern works best with non-overlapping starting events. It is problematic to associate a unique ending event with multiple unique starting events, unless each event has a unique ID or tag associated with it. In section 6.4.2 "Sequence implication pattern" on page 181 and section 6.7.4 "Pipelined protocol pattern" on page 199 we demonstrate how to handle overlapping transactions.

6.7 Sequence patterns

This section presents patterns involving sequences of Boolean expressions that span across multiple cycles. These patterns typically describe protocols (implied or expressed) or FSM transitions. They include *forbidden sequence*, *comparing captured data*, and *tagged transaction* patterns.

6.7.1 Forbidden sequence patterns

Pattern name. Forbidden sequence

Problem. Simple two-cycle (previous and current or current and next) combinations of signals should not occur within a design. Often there are multi-cycle combinations of events that can not occur.

Motivation. Consider a cache protocol with the following requirement: whenever an active `invalidate` occurs, then another cache `invalidate` or `hit` must never occur within the next four cycles. Often, it is easier to specify a forbidden sequence than specify all legal sequences.

Context. The following are example situations that provide opportunities to detect incompatible conditions that may lead to incorrect operations:

- Control logic that creates next cycle inputs that are illegal in specific states (flushing, invalidating, stalls, and so forth)
- Protocols that note illegal combinations of signals across a number of cycles
- State machines that have inputs that are illegal for a given relationship between current and next states

Solution. Use a forbidden sequence to identify the illegal occurrence of a `hit` and `invalidate` in the quiescent period.

Forbidden sequences allow a variable or fixed-width specification to be applied to expressions. The variable width specification allows us to use one sequence to account for the four cycles where a *hit* is not expected:

Example 6-39 ***PSL*** **forbidden sequence check**

```
// declarative assertion
// Once an invalidate occurs, neither hit or invalidate may be
// asserted for the next 4 cycles.

assert never {invalidate; {invalidate || hit} [*1:4] } @(posedge clk);
```

Considerations. Forbidden sequences can be specified using the OVL assert_cycle_sequence. For example, to specify that the sequence `A` followed by `B`, which is then followed by `C` should never occur, we can specify the forbidden sequence as a Verilog concatenation as follows:

```
{A, B, C, 1'b0}
```

If the prefix sequence is match (that is, `{A, B, C}`), then the assert_cycle_sequence will flag a failure since the last element in the sequence is defined as `1'b0`.

6.7.2 Buffered data validity pattern

Pattern name. Buffered data validity

Problem. Data can be dropped (that is, lost) or corrupted when attempting to transfer data across interfaces that involve latency.

Motivation. When transferring data between blocks on a shared bus, there is a risk of dropping data during a transfer. Hence, it is necessary to ensure that the received data matches the transferred data.

Context. This pattern is useful for protocols that transfer information (such as, data and address). It allows for assertions that describe correct operation without requiring additional RTL to capture the data.

Solution. The assertion in Example 6-40 captures the transfer data at the beginning of the transaction and compares the captured data with the received data at the end of the transfer.

Example 6-40 ***SystemVerilog captured data* check**

```
// declarative assertion
// Capture data (into tdata) and compare it when you see the new
// transaction (new_trans).

property capture_check;
  reg [31:0] tdata;
  @(posedge clk) (new_req, tdata=data |-> ##[1:100] new_trans
      ##0 tdata == trans_cmd);
endproperty

assert property (capture_check)
  else $error ("Transaction (%0d) not started within 100 cycles, or trans_cmd (%0d) wrong.",
         new_trans, trans_cmd);
```

Considerations. This pattern applies to various control and datapath sections of logic. However, for some protocols, this pattern is insufficient. This is the case for protocols implementing pipelined transactions. See also Section 6.7.3 "Tagged transaction pattern" and Section 6.7.4 "Pipelined protocol pattern", which discuss techniques to handle overlapping transactions.

6.7.3 Tagged transaction pattern

Pattern name. Tagged transaction

Problem. Ensure that transactions consisting of multiple out-of-order responses match the appropriate initial request.

Motivation. To improve throughput, many protocols allow multiple transactions to complete in an out-of-order fashion. For example, a tag (that is, unique ID) is often associated with a transaction's initial request, while another tag is associated with the transaction's ending response. To ensure that no data is lost while processing the transaction, it is necessary to validate that for any given tag associated with an initial request, there eventually exists an ending tag with the same ID value.

Consider the interface protocol illustrated in Figure 6-14. This interface can initiate a transaction with an active `req` and an accompanying four-bit request `tag` that contains a unique transaction ID. The transaction completes when `done` is asserted and the done `tag` equals the same value as the request `tag`.

Figure 6-14 **Tagged transaction.**

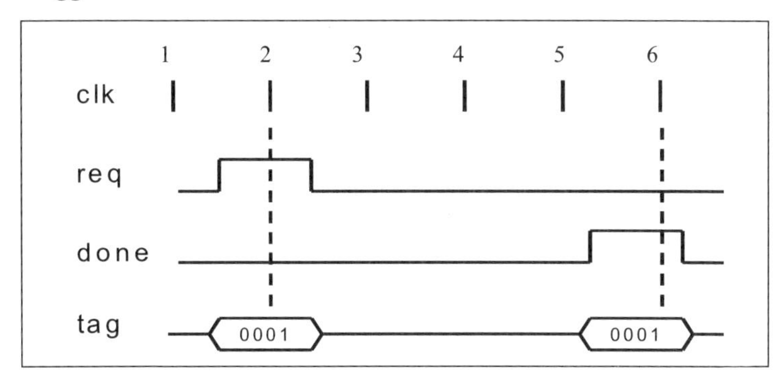

Context. The *tagged transaction* pattern is useful in bus protocols when there is a potential for different latencies to exist between multiple bus components, which results in an out-of-order response.

Solution. Validate the proper completion of a tagged transaction by creating a set of assertions that meet the following requirements:

- Ensure that a transaction is never lost (that is, completes within 100 cycles).
- Ensure that a unique tag is never reused (that is, multiple *requests* or *completions* for the same tagged transaction are not allowed).
- Ensure that there is one completion for each request.

Example 6-41 demonstrates how to code these assertions in SystemVerilog.

For an out-of-order transaction, a request tag must not be reused until after the original request (with the same tag) completes. Thus, our second property (`reqtag_once`) in Example 6-41 specifies that another request will not occur with the same tag before the acknowledge (`done`) is returned for the original request with the same tag. It is also required that only a single acknowledge is returned for a given tag. Hence, we use this same technique to check this condition. That is, when `done` is received, we specify that an additional `done` responses cannot be received for the same requested tag.

Example 6-41 *SystemVerilog* tagged `req/ack` protocol

```
  . . .

// Property definitions.
// assert request 1 completes within 100 cycles.
property req2done;
  int rtag;
  @(posedge clk) (req, rtag=req_tag |->
     ##[1:100] done && done_tag == rtag);
endproperty

property reqtag_once;
  int rtag;
  // Once a request (with a specific tag) is made,
  // there must be a done with that same tag, before
  // another request with the same tag is issued.
  @(posedge clk) not (req, rtag=req_tag
     ##1 !(done && done_tag == rtag) [* 1:$]
     ##0 req && req_tag == rtag);
endproperty

property donetag_once;
  int dtag;
  // Once a done (with a specific tag) is issued,
  // there must be a request with that same tag, before
  // another done with the same tag is issued.
   @(posedge clk) not (done, dtag=done_tag
     ##1 !(req && req_tag == dtag) [* 1:$]
     ##0 done && done_tag == dtag);
endproperty

// Concurrent assertion statements.
assert (req2done)
    else $error("Request tag didn't complete within 100 cycles.");

assert property (reqtag_once)
    else $error("Request tag re-used before done received.");

assert property (donetag_once)
    else $error("Two acknowledges received for request tag.");
```

replicating a set of assertions

The forall construct in PSL conveniently enables us to replicate a set of assertions related to a unique tag as shown in Example 6-42.

In other words, the forall construct in Example 6-42 creates eight unique assertions associated with each tag, which are then used to validate the completion of the transaction within 100 cycles, and to validate that a given request tag will not be reused within 100 cycles.

Figure 6-15 **Done tag 0000 transaction error**

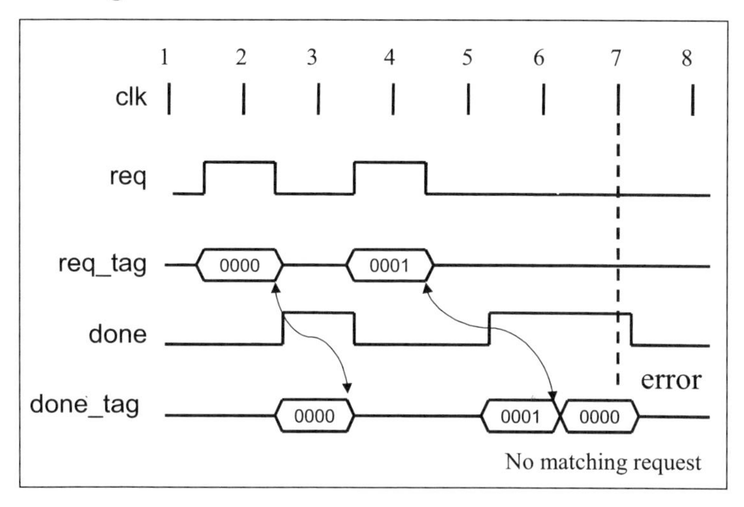

Example 6-42 ***PSL* multiple tagged `req/ack` protocol**

```
// declarative assertion
// assert any request 'n' completes (receives its done) within 100
// cycles.

assert forall T in {0:7} :
  always req && tag==T -> next {[*1:99]; done && done_tag==T}
    @(posedge clk);

// Note this assertion does not ensure that there is only 1 done
// corresponding to the original request.
```

Considerations. The tagged transaction pattern is not appropriate for all protocols. For inorder protocols, where a given transaction must complete prior to the completion of the next transaction, apply the pipelined protocol pattern discussed in Section 6.7.4 "Pipelined protocol pattern".

6.7.4 Pipelined protocol pattern

Pattern name. Pipelined protocols.

Problem. Due to the overlapping nature of inorder transactions, and the implicit ordering, it is difficult for assertions alone to associate a specific transaction completion event with the corresponding request. Hence, more than one request can falsely match a single transaction completion event.

Motivation. Pipelined protocols allow for greater throughput on an interface by allowing a limited number of additional transactions to begin before previous transactions complete. Yet, matching the appropriate transaction completion event with an appropriate request is problematic.

Consider a simple inorder handshake protocol for servicing a request. A `request` signal starts a transaction. An `acknowledge` signal completes the transaction. In general, there is a restriction on the maximum number of outstanding requests that are supported by the protocol. For this protocol, we require that each acknowledge completes the oldest initiated transaction. This strict ordering of transactions must be maintained, as illustrated in Figure 6-16.

Figure 6-16 **REQ/ACK with FIFO in order semantics**

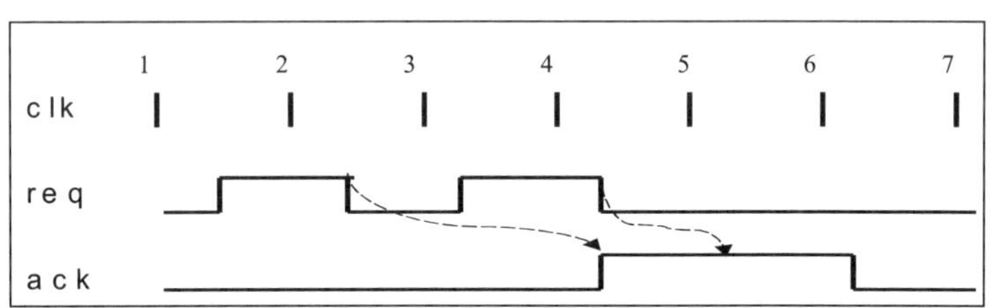

Context. Pipelined (inorder) protocols are common on design interfaces as well as within control logic. Queue (FIFO) modules implicitly have this protocol, since they store and output (*push* and *pop*) data in the same order it comes in. The queue depth represents the maximum number of outstanding requests that the protocol can support.

Example 6-43 ***PSL* and Verilog pipelined `req/ack` handshake protocol**

```
// Setup two counters to tag the req's and ack's.
reg [3:0] req_cnt, ack_cnt;
initial {req_cnt, ack_cnt} = 8'b0;

always @(posedge clk) begin
  // Increment counter each time event is seen.
  if (req) req_cnt <= req_cnt + 1;
  if (ack) ack_cnt <= ack_cnt + 1;
end

// declarative PSL assertion
// assert tagged req/ack sequence completes in 100 cycles

// For each tag.
assert forall C in {0:15} :
  always
   {req && req_cnt == C} |=>  {[*0:99]; ack && ack_cnt == C}
      @(posedge clk);
```

Example 6-44 *SystemVerilog* **pipelined_reqack module**

```
module pipelined_reqack // all arguments are inputs
  (         clk,
            req,
            ack,
            req_datain, // Used to sample data at request time.
            dataout,    // Data exiting fifo to compare.
            latency,   // a constant expression
            pipedepth
   );
  // Note: the dataout input argument above is expected to be the data
  // exiting the FIFO. It will be compared with the stored data in
  // the assertion below. Since the parameter (and req_datain) have
  // default values of 0, their use is optional.

  // Setup two counters to tag the req's and ack's.
  reg [3:0] req_cnt = 4'b0,
            ack_cnt = 4'b0;

  // Update counters when req or ack event seen.
  always @(posedge clk)
    begin
      if (req) req_cnt <= req_cnt + 1'b1;
      if (ack) ack_cnt <= ack_cnt + 1'b1;
    end

  property req_no_ack;
     int cnt, req_data; // dynamic variables
     @(posedge clk)
       (req, cnt=req_cnt, req_data=req_datain |->
          ##[1:latency] ack && ack_cnt==cnt
          ##0 req_data==dataout);
  endproperty

  property exceeded_depth;
    @(posedge clk)  (req_cnt - ack_cnt) < pipedepth);
  endproperty

  // We don't want to see an ack twice with the same tag without
  // a request with that tag between them.
  property no_extra_ack;
    reg [3:0] ackcnt;  // dynamic variables
    @(posedge clk) not (ack, ackcnt=ack_cnt
     ##1 !(req && req_cnt == ackcnt) [* 1:$]
     ##0 ack && ackcnt == ack_cnt);
  endproperty

  assert property (req_no_ack)
    else $error("Ack not received within timeout limits %0d or ack_check_seq failed.",
         latency);
  assert property (exceeded_depth)
    else $error("Exceeded pipedepth of interface.");
  assert property (no_extra_ack)
    else $error("Additional ack not matching req.");
endmodule
```

Solution. Apply the *pipelined protocol pattern* to a design to ensure the proper completion of `req-ack` pairs as demonstrated in Example 6-43. For this example, we create two counters `req_cnt` and `ack_cnt`, used to 'tag' each request and acknowledge with a matching number. This enables us to support up to sixteen outstanding requests. At every `req` and `ack` event, we increment the appropriate counter. The assertion we write specifies that for all active `req` (and the associated request count) there is an active `ack` whose acknowledge count matches the request count. Note that this example can support up to 16 outstanding requests.

Considerations. This *pipeline protocol pattern* can be extended to check transferred data. For example, if you are checking a FIFO, you can validate that data placed into a FIFO is read out in the correct order. In Example 6-44 we demonstrate how to code a SystemVerilog module, which can be instantiated in the SystemVerilog RTL to validate correct ordering of data.

This module is not limited to validating FIFOs. Due to its storage capabilities, it is also useful for validating a bus interface that permits overlapping transactions and can be used on blocks that buffer transfer data. This is useful for designs where the output transmission time is unpredictable due to the variety of events.

6.8 Applying patterns to a real example

The patterns we discussed in the previous sections provide multiple examples for effectively applying assertions. In this section, we explore an *assertion-based design* process with a real example, an SRAM controller. The design process begins with the engineer reviewing a design specification and refining the set of requirements into an RTL model that is ultimately synthesized into a gate-level implementation.

Let's consider the Verilog-2001 module interface shown in Example 6-45:

Methodology guideline

The best process for adding assertions to a module is the following:

- Add assertions to each interface of a module. These assertions help to define the interface protocol, legal values, required sequencing, and so forth.
- Add assertions between interfaces of a module. These assertions help to define how the module operates on information from its interfaces and what it is supposed to do.
- Add assertions as you code structures within your module defining design intent (that is, acceptable operating

conditions). This will identify simple mistakes due to incorrect internal operations.

- Add assertions to the control logic you implement that ties the interfaces, structures, and remaining logic.

Example 6-45 SRAM module interface definition

```
module sramInterface
 #(
   parameter ADDRW = 16,  // Number of address bits.
   parameter QueDepth = 6 // 2 <= QueDepth <= 16
  )
 (
  localparam IWB = ADDRW-2; // Upper address bit.
/*
  The abbreviated specification of this sram controller is the
  following:
  Provide a queued interface to a sram that accepts requests for
  reads/writes, but will not accept writes when previous reads are
  queued. Requests are issued to memory and completed when the memory
  sends memory done.
*/
input           clk;       // The clock
input           rst_n;     // The reset

// SM interface to Queue.
input           SMQueNew;  // New request on SM interface.
input  [IWB:1]  SMQueAddr; // Address of request
input           SMQueSpec; // speculative request (only with reads).
input  [ 3:0]   SMQueWrEn; // Write enables (0000,0011,1100,1111 valid)
input  [31:0]   SlvWrData; // Data written with enables (each byte)

output [IWB:2]  QueBufAddr; // Address to be completed on ENG interface
output          QueBufValid;// Read request done.

// ENG interface from Queue to memory.
output          EngMemRd;    // Valid read request.
output          EngMemWr;    // Valid write request.
output [IWB:2]  EngMemAddr;  // ENG address for operation.
output [63:0]   EngMemData;  // ENG write data
output [ 3: 0]  EngMemWrEn;  // ENG write enables (each word).
input           MemEngDone;  // Request is complete from memory.

// Queue status bits.
output          QueAlmostFull;    // One more entry available.
output          QueFull;          // No more entries can be accepted.
output         ReadExistsInQue;  // Writes not sent to Q until no reads.

// Queue management interface.
input          Flush;        // Hold flush to flush the Q. Accept no new
output         FlushAck;     // until FlushAck asserted-then clear Flush
input           SMQueStop;    // Hold Stop to empty Q.
output   reg    QueSMStopAck;// Ack will occur when Stop and Q empty.

endmodule
```

In Section 6.8.1 "Intra-interface assertions" we discuss and provide example assertions that apply *to* each of four interface types. Then in Section 6.8.2, "Inter-interface assertions", we offer assertion examples that apply *across* interfaces to validate that the block is performing correctly. Each example includes the specific pattern name we used above and offers an application that solves particular assertion requirements.

6.8.1 Intra-interface assertions

We begin by discussing intra-interface assertions. These apply to the following four interface types:

An interface is a set of signals that implement a protocol or transmit information

- New request interface
- SRAM interface
- Queue status interface.
- Queue management interface.

6.8.1.1 New request interface

The *new request interface* defines a valid signal (`SMQueNew`) and several data signals that define the request. Write the following assertions for this purpose:

Example 6-46 ***SystemVerilog*** **During a request, all signals must evaluate to 1 or 0**

```
// X detection pattern

assert property (@(posedge clk) disable iff ( rst_n )
   not (SMQueNew |->
            $isunknown(SMQueAddr, SmQueSpec, SMQueWrEn, SlvWrData)))
  else $error("Unknown signal value when asserting a new request.");
```

Example 6-47 ***SystemVerilog*** `SMQueSpec` **must not be asserted during a write request**

```
// invalid signal combination pattern

assert property (@(posedge clk) disable iff ( rst_n )
    not (SMQueNew & SmQueWrEn & SmQueSpec))
  else $error("Received illegal request for speculative (SMQueSpec) write.");
```

Example 6-48 *SystemVerilog* Only certain enable patterns allowed during write request

```
// conditional pattern

assert property (@(posedge clk) disable iff ( rst_n )
  (SMQueNew |->
     $insetz(SMQueWrEn[3:0], 4'b0000, 4'b0011, 4'b1100, 4'b1111)))
  else $error("SmQueWrEn has illegal write enable pattern %b.",
           SmQueWrEn);
```

6.8.1.2 SRAM interface

The SRAM interface also specifies two signals (`EngMemRd` and `EngMemWr`) that validate the request. Assertions for this interface are:

Example 6-49 *SystemVerilog* Unknown signals not allowed during a valid `EngMemRd` or `EngMemWr` request

```
// X detection pattern

assert property (@(posedge clk) disable iff ( rst_n )
    not (EngMemRd | EngMemWr |->
          $isunknown(EngMemAddr, EngMemData, EngMemWrEn)))
  else $error("Sram interface contains unknown values.");
```

Example 6-50 *SystemVerilog* `EngMemRd` and `EngMemWr` are mutually exclusive

```
// valid signal combination pattern

assert property (@(posedge clk) disable iff ( rst_n )
    not (EngMemRd & EngMemWr)) else
  $error("Sram interface asserts illegally read and write together.");
```

Example 6-51 *SystemVerilog* For a valid write, the `EngMemWrEn` may only have one of seven legal values

```
// valid combination of signals pattern

assert property (@(posedge clk) disable iff ( rst_n )
    (EngMemWr |-> $inset(EngMemWrEn,0,1,2,3,4,8,12))) else
  $error("Sram interface asserts illegal write en (%0b)", EngMemWrEn);
```

Example 6-52 ***SystemVerilog*** **For a request, the completion signal** `MemEngDone` **must eventually be returned**

```
// pipeline protocol pattern
// See pipelined_reqack templete definition in Example 6-44

// Make sure queued read is completed by memory (with MemEngDone).
// Atmost 6 requests (pipedepth) may be queued.
// Limit latency for return to 100 cycles.

pipelined_reqack
   sendReadReq(.req(EngMemRd), .req_datain(),
               .ack(MemEngDone),
               .dataout(),
               .clk(clk), .latency(100), .pipedepth(6));
```

The assertion in Example 6-52 helps define the interface by specifying that the module can send a new request during the same cycle that `MemEngDone` is returned (to decrease latency). By using the assertion to show that the overlap is possible, designers get more of the type of information that is necessary to complete the design.

6.8.1.3 Queue status interface

The queue status interface relays the state of the queue to the interfacing modules. Assertions for this interface are:

Example 6-53 ***SystemVerilog*** `QueFull` **and** `QueAlmostFull` **are mutually exclusive**

```
// valid signal combination pattern

assert property (@(posedge clk) disable iff ( rst_n )
     not (QueFull & QueAlmostFull))
   else $error("Queue interface illegal state Quefull and QueAlmostFull.");
```

Example 6-54 ***SystemVerilog*** **The queue status must not contain unknown values**

```
// X detection pattern

assert property (@(posedge clk) disable iff ( rst_n )
     $isunknown(QueFull, QueAlmostFull, ReadExistsInQue))
   else $error("Queue status interface has unknown values.");
```

6.8.1.4 Queue management interface

The queue management interface gives control to the other blocks to direct the shutdown of the interfaces. Assertions for this interface are:

Example 6-55 ***SystemVerilog* A `Flush` must remain active until `FlushAck` is asserted**

```
// event-bounded window pattern

assert property (@(posedge clk) disable iff ( rst_n )
    ($rose(Flush) |-> (Flush throughout (##[0:100] FlushAck)))) else
  $error("Flush must be held until Ack asserted.");
```

Example 6-56 ***SystemVerilog* `FlushAck` must be asserted for no more than one cycle**

```
// forbidden sequence pattern

assert property (@(posedge clk) disable iff ( rst_n )
    not (FlushAck ##1 FlushAck)) else
  $error("FlushAck must be a single pulse.");
```

Example 6-57 ***SystemVerilog* A `SMQueStop` must remain active until `StopAck` is asserted**

```
// event-bounded window pattern

assert property (@(posedge clk) disable iff ( rst_n )
     ($rose(SMQueStop) |->
         (SmQueStop throughout ##[0:100] StopAck)))
   else $error("Stop must be held until ack is asserted.");
```

Example 6-58 ***SystemVerilog* `StopAck` must be asserted for no more than one cycle**

```
// forbidden sequence

assert property (@(posedge clk) disable iff ( rst n )
    not (StopAck ##1 StopAck)) else
  $error("StopAck must be a single pulse.");
```

Example 6-59 ***SystemVerilog* `FlushAck` and `!Flush` are mutually exclusive**

```
// valid signal combination pattern

assert property (@(posedge clk) disable iff ( rst_n )
    not (FlushAck & !Flush)) else
  $error("FlushAck must not assert without Flush asserted.");
```

Example 6-60 *SystemVerilog* **When a** `Flush` **is requested, the** `Flush` **must be acknowledged within 100 cycles**

```
// time-bounded combination pattern

assert property (@(posedge clk) disable iff ( rst_n )
    ($rose(Flush) |-> ##[1:100] IntFlushAck))
  else $error("Flush took too long (>100 cycles) to complete.");
```

The assertions shown above are applied to individual interfaces to a block. Next, we'll look at assertions that are written *across* interfaces.

6.8.2 Inter-interface assertions

Since an interface is typically driven from a single block, errors on the interface can be directly attributed to this one source. However, the assertions in this section comprise a set of inter-interface assertions. They interact with several interfaces to further check for correct operation or check for proper operation of the block given the data received. The assertions are:

Example 6-61 *SystemVerilog* **When there is a new valid read request, the address must occur on the SRAM interface within 100 cycles, and the SRAM address must match the request address**

```
// pipelined-protocol pattern

// See pipelined_reqack templete definition in Example 6-44

pipelined_reqack
   sendReadReq(.req(SMQueNew & !SMQueWrEn), .valin(SmQueAddr),
               .ack(EngMemRd),
               .dataout(EngMemAddr),
               .clk(clk), .latency(100), .pipedepth(6));
```

Example 6-62 *SystemVerilog* **When there is a new valid read request, there must be a completion of this read** `QueBufValid`

```
// pipelined-protocol pattern

// See pipelined_reqack templete definition in Example 6-44

pipelined_reqack
   receiveData (.req(SmQueNew & !SMQueWrEn), .req_datain(),
               .ack(QueBufValid),
               .dataout(),
               .clk(clk), .latency(100), .pipedepth(6));
```

Example 6-63 *SystemVerilog* **When there is a new valid write request, the address and data must occur on the SRAM interface within 100 cycles, and the SRAM address and data must match the request**

```
// pipelined-protocol pattern

// See pipelined_reqack module definition in Example 6-44

pipelined_reqack
   sendWriteReq(.req(SMQueNew & |SMQueWrEn),
                .req_datain({SlaveWrData, SmQueAddr}),
                .ack(EngMemWr),
                .dataout({EngMemData, EndMemAddr}),
                .clk(clk), .latency(100), .pipedepth(6));
```

Example 6-64 *SystemVerilog* **Check that when a new valid read occurs,** `ReadExistsInQue` **is asserted the next cycle**

```
// conditional pattern

assert property (@(posedge clk) disable iff ( rst_n )
      (SmQueNew & !SmQueWrEn |-> ReadExistsInQue)) else
  $error("ReadExistsInQue not asserted for valid read.");
```

Example 6-65 *SystemVerilog* **If asserting** `SMQueStop` **or** `Flush` **(internal flush state), no new requests can be received**

```
// conditional pattern

assert property (@(posedge clk) disable iff ( rst_n )
    ((Flush | SMQueStop) |-> SMQueNew))
  else $error("Illegal SMQueNew while asserting Stop or Flush.");
```

Example 6-66 *SystemVerilog* **A new valid write must not be asserted while** `ReadExistsInQue`

```
// valid signal combination pattern

assert property (@(posedge clk) disable iff ( rst_n )
    not (SmQueNew & |SMQueWrEn & ReadExistsInQue)) else
  $error("%s\n%s", "A write is being enqueued while there is a valid read in the queue.",
    "This must be avoided so that the read can be invalidated by the write.");
```

Example 6-67 *SystemVerilog* **If** `QueAlmostFull` **and a new valid read request and no completion of a request, then** `QueFull` **must be asserted the next cycle**

```
// valid signal combination pattern

assert property (@(posedge clk) disable iff ( rst_n )
    (QueAlmostFull & !MemEngDone & SMQueNew |-> QueFull)) else
  $error("Queue did not fill when one more request entered.");
```

Example 6-68 ***SystemVerilog*** **`SmQueNew` and `QueFull` are mutually exclusive**

```
// valid signal combination pattern

assert property (@(posedge clk) disable iff ( rst_n )
    not (SmQueNew & QueFull)) else
  $error("Overflow of SMQue block.");
```

Together, the inter- and intra-interface assertions provide a valuable net for trapping incorrect behavior for a block. They not only detect illegal operation of the block, which is valuable for allowing quick corrections, they also detect any illegal stimulus from verification testcases, testbench stimulus generators, or the actual companion system block. The complete process includes adding assertions to the implementation of this block, including the necessary structures, and the derived logic to the interfaces of the structures.

6.9 Summary

In this chapter, we introduced the concept of an *assertion pattern* as a convenient medium used to document and communicate design insight for assertions that recur. The goal in creating an *assertion pattern* is to present two interdependent components (that is, *property structure* and *rationale*) about a particular characteristic or verification concern associated with a particular design.

The patterns we introduced in this chapter include:

- Signal patterns
- Set patterns
- Conditional patterns
- Past and future event patterns
- Window patterns
- Sequence pattern

CHAPTER

7

ASSERTION COOKBOOK

Like a culinary cookbook that offers new ways to prepare traditional dishes, this chapter offers a set of recipes for cooking up assertions and functional coverage for many common design structures found in today's RTL designs. Just as a good cookbook offers more than lists of ingredients, the goal of our cookbook is to offer examples of design assertions and functional coverage coding techniques that combine the right ingredients and methods to achieve successful coverage. As you consider applying our examples to your own designs, we recommend that you initially write assertions that specify the design intent prior to coding the RTL implementation. Capture additional assertions during the process of coding the RTL implementation as design details develop. Also, specify implementation functional coverage points during RTL coding to ensure that key features of the lower-level implementation are adequately tested—along with the higher-level functionality and functional coverage defined by the verification team.

In this chapter, we explore a typical set of assertions and functional coverage points for queues, stacks, finite state machines, encoders, decoders, multiplexers, state table structures, memory, and arbiters. Certainly our list of common structures is not exhaustive. Nonetheless, if you combine these common examples with the patterns examples discussed in Chapter 6, you can extend these ideas to cover your own unique designs.

add functional coverage to RTL only when it makes sense

Many of our examples demonstrate how to specify functional coverage for various design structures. However, we advise you to selectively add RTL implementation functional coverage. It is only for portions of your RTL implementation that concerns you—specifically related to adequate testing (such as boundary conditions on queue pointers, which can be hard to observe or forgotten, in a chip-level verification environment). Some of our examples, such as functional coverage on multiplexer select lines or counter values and controls, could generate enough data to overwhelm the engineer analyzing the results. In addition, for some of the lower-level functional coverage examples (again, for example the multiplexer select lines), the coverage could be measured using a commercial code coverage tool. However, our goal in this chapter is to demonstrate different forms of specification using the structures described in this section, you must decide what makes sense specific to your testing goals when specifying functional coverage in your own RTL implementation. See section 5.4.2 "Correct coverage density" on page 137 and section 5.4.3 "Incorrect coverage density" on page 139 for detailed discussions of appropriate specification of functional coverage. Incidentally, some functional coverage tools have an upper limit on the reporting of a particular functional coverage point. This is useful to minimize the noise that can occur during verification when incorrectly specifying functional coverage on high occurrence events.

user-defined error messages

For many of the examples in this section, we have omitted *user-defined error messages* and property names for simplification. However, we recommend that you always code user-defined error messages for your assertions (assuming your assertion language permits reporting user error messages), since doing so provides a context for other users who analyze simulation failures and do not know the specific details of the design (or assertion). Refer to section 2.2.2 "Best practices" on page 33.

clocks and resets

For simplicity, many of our PSL examples are coded under the assumption that the engineer had previously defined a default clock in PSL. For example:

default clock = (**posedge** clk);

Furthermore, many of our examples, for simplicity, do not take into account a reset condition. We recommend that you augment our examples with either the PSL abort:

assert always (a^b) @(**posedge** clk) **abort** rst_n;

or SystemVerilog disable iff operator to handle your specific reset requirements:

assert property (@(**posedge** clk) **disable iff** (rst_n) (a^b));

7.1 Queue—FIFO

Context. The *queue* (that is, FIFO), one of the most easily distinguished elements in a design, is primarily used to buffer data between multiple processing elements—typically operating at dissimilar rates. Unfortunately, there are multiple implementations of queues containing many unique features, even within the same design. However, there are several common features in all queue structures that warrant assertions. By specifying assertions associated with common features, you can validate that data is neither lost nor used incorrectly. Specifically, the control circuits designed to prevent overflow or underflow conditions are ideal queue features that assertions should check.

Example 7-1 and Example 7-2 are common implementations of a queue. In Example 7-1, we implement a queue with an index counter (which represents the current depth of the queue). Alternatively, Example 7-2 implements a queue using a unique read and write pointer.

In the following sections, we demonstrate a number of common assertions associated with queues. These assertions detect the error conditions such as: *overflow, underflow, invalid status, data corruption, invalid flush,* and *invalid state.* In addition, we recommend a number of likely features you might wish to specify as functional coverage.

Example 7-1 ***Verilog*** **fragment for FIFO with index counter**

```
// use a counter to represent current depth of the FIFO
parameter FIFO_depth=16;
parameter logDEPTH=4;
reg [logDEPTH-1:0] cnt;

always @(posedge clk) begin
  case ({push, pop})
    2'b10: begin
             cnt <= cnt + 1;
             ...
    2'b01: begin
             cnt <= cnt - 1;
             ...
```

Example 7-2 *Verilog* **fragment for FIFO using read and write pointers**

```
// use two decoded pointers starting with value 1
// when pointers are equal (and full is not set) FIFO is empty
parameter FIFO_depth=16;
reg [FIFO_depth-1:0] rdptr, wrptr;
reg                   full;
reg                   empty;

always @(posedge clk or negedge rst_n) begin
  if (!rst_n) begin
    rdptr <= 1;
    wrptr <= 1;
    full  <= 0;
    ...
  end
  case ({push, pop})
    2'b10 : // WRITE
     begin
  // rotate the write pointer, set full if update will match rdptr
      wrptr <= {wrptr[FIFO_depth-2:0], wrptr[FIFO_depth-1]};
      full  <= rdptr == {wrptr[FIFO_depth-2:0], wrptr[FIFO_depth-1]};
       ...
    2'b01 : // READ
     begin
  // rotate read pointer, clear full since we can't
      rdptr <= {rdptr[FIFO_depth-2:0], rdptr[FIFO_depth-1]};
      full <= 1'b0;
       ...
```

7.1.1 Assertions

7.1.1.1 FIFO overflow assertion

detecting overflow

Pattern name. FIFO overflow

Problem. Overflowing a FIFO causes data loss.

Solution. The assertions in Example 7-3 specify that it is not possible to write a new element into the FIFO when the FIFO is full. Note that these assertions are a variation on the *invalid signal combination pattern* discussed in section 6.3.3 on page 177.

Example 7-3 *PSL* **FIFO overflow assertion**

```
assert never (cnt == FIFO_depth & push & ! pop); // for Example 7-1
assert never (full & push & !pop);               // for Example 7-2
```

Regardless of the two implementations, both assertions share the same pattern. That is, it is invalid to write into the FIFO while the FIFO is full.

7.1.1.2 FIFO underflow assertion

detecting underflow

Pattern name. FIFO underflow

Problem. Invalid data read from an empty FIFO results in unpredictable behavior, which can be difficult to debug and isolate at the chip-level during verification.

Motivation. Similar to the overflow problem, attempting to remove too much data creates unpredictable results (that is, an error in the design). When you specify an assertion for an underflow situation, you must take into account the architecture of the FIFO. For example, for a feed forward FIFO, the input data port is directly routed to the output data port when the FIFO is in an empty state.

Solution. Example 7-4 specifies that it is not possible for the design's control logic to read from an empty FIFO. Hence, this assertion identifies underflow conditions during verification.

Example 7-4 ***PSL*** **FIFO underflow with feed forward design**

```
assert never ((cnt == 0) & !push & pop);                  // Example 7-1
assert never ((wrptr == rdptr) & !full & !push & pop); // Example 7-2
```

Example 7-5 demonstrates how to specify an underflow assertion for a non-feed forward FIFO. That is, the FIFO does not allow direct routing of the input data to an output data port. Hence, the push signal is not part of the assertion.

Example 7-5 ***PSL*** **FIFO underflow without feed forward**

```
assert never ((cnt == 0) & pop);                  // for Example 7-1
assert never ((wptr == rdptr) & !full & pop);  // for Example 7-2
```

7.1.1.3 FIFO non-corrupt data assertion

ensure that outgoing data matches original incoming data

Pattern name. FIFO non-corrupt data

Problem. If the data leaving the FIFO doesn't match the original data entering the FIFO (with the appropriate latency), then the read data is corrupt and can result in unpredictable behavior, which can be difficult to debug and isolate at the chip-level during verification.

Motivation. Detecting errors within a FIFO design is easiest when you identify them close to the source and time of occurrence. Often you must track bus transaction errors back to a specific FIFO because the data is not correct. By using FIFO non-corrupt data assertions, you will be able to accurately identify problems, such as FIFO flush, complicated push/pop operations, and general data corruption.

Solution. Example 7-6 uses a module to ensure that the final data read out of the FIFO is not corrupted with respect to the original data written into the FIFO. This module makes use of the new SystemVerilog *dynamic variable* data facilities associated with properties and sequences, which enables us to store incoming FIFO data to compare with the output data of the FIFO at a future point in time. For a more complete definition of the pipeline_reqack module, see Example 6-44, "SystemVerilog pipelined_reqack module" on page 201.

Consideration. The pipeline_reqack module instantiation in Example 7-6 is connected to the FIFO control signals `push` and `pop`. The incoming data (that is, `data_in`) is stored in the FIFO for comparison when a `push` occurs. Later, the stored data is compared with the `dataout` value when a `pop` occurs.

Example 7-6 ***SystemVerilog*** **module to ensure FIFO is not corrupted**

```
// see details section Example 6-44 on page 201
 pipeline_reqack FIFO_check
           (.pipedepth  (8),          // depth of FIFO
            .latency    (20),         // latency to acknowledge
            .req        (push),
            .req_datain (data_in),    // data put into FIFO
            .ack        (pop),
            .dataout    (data_out),   // data exiting the FIFO
            .clk        (clk));
```

7.1.1.4 FIFO flush assertion

ensure that a FIFO flush operates correctly

Pattern name. FIFO flush

Problem. If FIFO data is not invalidated during a FIFO flush, and is read at some future point in time, then unpredictable behavior can occur, which can be difficult to debug and isolate at the chip-level during verification.

Motivation. One source of FIFO design errors is associated with FIFO flushing (or invalidation) logic. Hence, we must ensure that all data is invalidated after a flush operation.

Solution. Consider the requirement for a FIFO containing a `flush` signal that when active, invalidates the FIFO data on the cycle immediately after the active flush. For the FIFO in Example 7-7, we could write an OVL assertion to ensure that all counters and read and write pointers have been properly reset after a flush.

Example 7-7 ***OVL* valid flush operation**

```
// for Example 7-1
assert_next flush_7_1 (clk, rst_n, flush, (cnt == 0));
// for Example 7-2
assert_next flush_7_2 (clk, rst_n, flush, (!full && rdptr == wrptr));
```

7.1.1.5 FIFO contiguous data assertion

ensure legal pointer state

Pattern name. FIFO contiguous data

Problem. If the FIFO valid bits, representing queued data, are not in a legal state, unpredictable behavior can occur, which can be difficult to debug and isolate at the chip-level during verification.

Motivation. Some FIFO implementations use a valid bit structure (as compared with a head/tail pointer implementation shown in Example 7-2) to track the element status. That is, one valid bit for each element in the FIFO indicates that the FIFO element is not empty. Under proper operation, the state of the FIFO must contain contiguous valid entries, which represent no holes in the FIFO data.

Solution. In Example 7-8 we check legal combinations of the *valid bit* structure using the new SystemVerilog $inset system function to ensure that data isn't lost.

Example 7-8 ***SystemVerilog* legal internal state for valid bits**

```
// Property for 6 deep FIFO to ensure legal state values
property legal_valid_states;
  @(posedge clk) ($inset(valid, 6'b0, 6'b1, 6'b11, 6'b111, 6'b1111,
                              6'b1_1111, 6'b11_1111));
endproperty
assert property (legal_valid_states);
```

In an alternate implementation shown in Example 7-9, we use a one-hot encoding for the read and write pointers. Each time a read or write occurs, the state rotates to the next available bit. We specify that the one-hot encoding is correct for all valid states. We also specify that all state transitions follow a valid rotation of encoding, which represents the contiguous content of the FIFO.

Example 7-9 ***SystemVerilog* valid pointer behavior**

```
parameter N=FIFO_depth;

property legal_ptr(ptr);
  @(posedge clk) ($countones(ptr)==1);
endproperty

property good_ptr_update (ptr);
  @(posedge clk)
    ($inset(ptr,
           0,                                 // reset
          $past(ptr),                         // stay same
           $past ({ptr[N-2:0], ptr[N-1]})));  // rotate
endproperty

assert property (legal_ptr(wrptr));
assert property (legal_ptr(rdprt));
assert property (good_ptr_update(wrptr));
assert property (good_ptr_update(rdptr));
```

In an alternate implementation shown in Example 7-9, we use a one-hot encoding for the read and write pointers. Each time a read or write occurs, the state rotates to the next available bit. We specify that the one-hot encoding is correct for all valid states. We also specify that all state transitions follow a valid rotation of encoding, which represents the contiguous content of the FIFO.

7.1.2 FIFO functional coverage

ensure proper coverage for FIFO full and empty cases

Pattern name. FIFO boundary condition

Problem. Without checking full and empty boundary conditions, key functionality could remain unverified.

Motivation. We want to ensure that the FIFO is sufficiently exercised. We could write functional coverage for each entry within the FIFO. Alternatively, we could focus on the end points (that is, FIFO full and empty conditions).

Solution. We use the full and empty status flags to determine the status of the FIFO. In Example 7-10 we show a functional coverage point using a PSL cover construct to report whenever the FIFO is full. We also report situations when the FIFO is empty.

Example 7-10 ***PSL* functional coverage for FIFO boundaries**

```
// for Example 7-2
default clock = (posedge clk);
cover { !full; full };
cover { !empty; empty };
```

7.2 Fixed depth pipeline register

Context. In addition to a FIFO operating as a buffer, FIFO pipeline structures are common in communication protocols across asynchronous clock domains. Pipeline registers often have stalling or flushing capabilities, which is typically where errors occur during implementation.

Figure 7-1 **Behavior of pipeline register of depth three with stall**

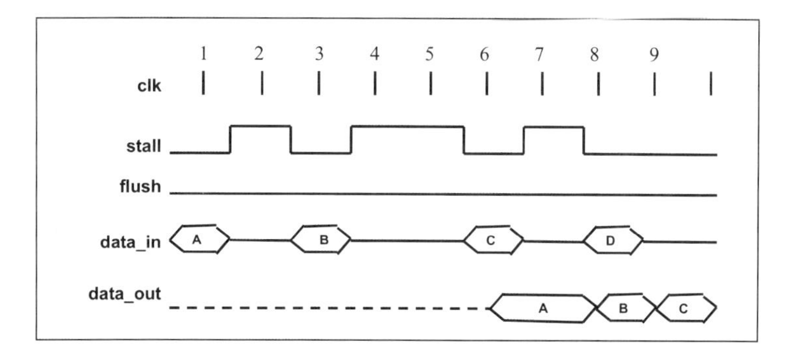

Example 7-11 demonstrates a simple pipeline register of depth three, with flushing and stalling capabilities. Figure 7-1 illustrates the behavior of the pipeline register. Data is read into the pipe when the `stall` command is inactive (and a `flush` is inactive). The data appears at the output of the pipe after two additional non-stalled cycles later. In addition, the current output is held constant during a `stall`.

Example 7-11 *Verilog* **fragment for simple pipeline register**

```
...
// width is the width of the data being piped
// depth is the number of pipe stages
parameter WIDTH=4;
parameter DEPTH=3;
parameter N = WIDTH*DEPTH; // size of register * # stages
reg [N-1:0] pipeline;

wire [WIDTH-1:0] data_in;
// output upper bits (WIDTH) of the pipe
wire [WIDTH-1:0] data_out = pipeline[N-1:N-WIDTH];

always @(posedge clk) begin
    if (flush)
      pipeline <= {N{1'b0}};
    else // shift value in pipe
      if (!stall)
        pipeline <= {pipeline[N-1-WIDTH:0], data_in[WIDTH-1:0]};
    end
end
```

7.2.1 Assertions

7.2.1.1 Pipeline data consistency assertion

ensure data is not corrupted through pipe

Pattern name. Pipeline data consistency

Problem. If a fixed depth pipeline register structure does not maintain consistency between the input data and the output data (accounting for the latency of the pipeline depth), then unpredictable behavior can occur, which can be difficult to debug and isolate during verification.

Motivation. The pipeline implementation may be a simple set of registers or it may be a more complex FIFO model with tags and multiplexers. For complex implementations where tags may be incorrectly flushed at the output of the pipeline, incorrect data can be returned due to an implementation error. Hence, it is critical to ensure that the pipeline register's data is consistent under various stalling and flushing scenarios.

Solution. In Example 7-12 we ensure that `data_in` entering the pipeline exits correctly to the `data_out` port. The SystemVerilog assertion uses the *dynamic variable* `thedata` to capture the data entering the pipe (that is, whenever a `stall` is not occurring). On the cycle immediately after the end of the sequence, where the number of *un-stalled* cycles within the sequence is equal to the depth of the pipe, the captured data `thedata` is compared with the data_out. Note that our implementation of the pipeline register is fairly straight forward. One could argue that the assertion we wrote is not useful for such a simple implementation—and we would agree. However, in many cases the pipeline register is not as simple as our example. Hence, our solution would apply to any implementation with stalling and flushing capabilities.

Example 7-12 ***SystemVerilog*** **pipeline data in—data out consistency check**

```
// related to Example 7-11
sequence seq_pipe_depth;
  // look for complete flow through pipe, including potential stalls,
  // without a flush
  !flush throughout !stall [*->DEPTH];
endsequence

// capture the data and wait until it exits the pipeline
// then compare against the saved value
property good_pipedata;
  reg [WIDTH-1:0] thedata;          // to save the incoming data
  // define a sequence that enables capturing data for comparison
  @(posedge clk)
    !stall, thedata=data_in       // first capture the data
    ##0 seq_pipe_depth            // valid transfer through pipe
    ##1 thedata == data_out;        // now check against the output
endproperty

assert property (good_pipedata);
```

7.2.1.2 Pipeline flush assertion

ensure pipeline data is flushed correctly

Pattern name. Pipeline flush

Problem. Failure to flush data in a pipeline can cause events (for example, a request) to occur earlier than expected, and the data itself will be out of sync with correct execution.

Motivation. Flushing pipeline registers may only require flushing valid signals, but in some cases data may also need to be flushed. The data that comes out incorrectly after a flush may cause incorrect control operations, or be data that should have been ignored.

Solution. The assertion we use in Example 7-13 ensures a correct flushing operation. Example 7-14 shows an assertion that checks that the data_out is flushed until valid data is clocked through the depth of the pipe.

Example 7-13 ***SystemVerilog* pipeline data is flushed**

```
// related to Example 7-11
property flushed_pipe;
  @(posedge clk)($rose(flush) |=> pipeline=={N{1'b0}});
endproperty

assert property (flushed_pipe);
```

Example 7-14 ***SystemVerilog* pipeline data is zero until pipeline is flushed**

```
// related to Example 7-11
property flushed_depth;
  @(posedge clk) ($fell(flush) |=>
    !data_out throughout !stall [*->DEPTH]);
endproperty

assert property (flushed_depth);
```

7.2.2 Functional coverage

ensure flushes occur during verification

Pattern name. Pipeline flush coverage

Problem. Without verifying that a flush occurs after writing into a pipeline register, key functionality could go unverified.

Motivation. Depending on the pipeline usage, coverage may include reporting the various data types sent through the pipe. For a general data pipeline, coverage of a proper flush operation is necessary.

Solution. As shown in Example 7-15, we report a special situation, which is all occurrences of a `flush` operation (and no `stall`) within two cycles of valid `data` appearing on the pipe output.

Example 7-15 ***SystemVerilog* functional coverage of flushed data**

```
// report occurrence of no stall and valid data followed
// within 1 or 2 cycles a flush
property cov_flushed;
  @(posedge clk)
  (!stall & |data ##[1:2] flush);
endproperty
cover (cov_flushed);
```

7.3 Stack—LIFO

Context. A stack is another common structure used for storing and retrieving data in a *last in first out* fashion, which is then used by a processing element. A stack is similar in complexity to a queue; and thus, so are the assertions you will write. Example 7-16 demonstrates a simple coding for a stack. In this example, we shift data into the variable stack_data whenever a push occurs, and we shift data out whenever a pop occurs. We maintain a set of valid bits for each stack entry to determine when the stack is full. In Section 7.1 "Queue—FIFO", we defined a set of assertions that are common to stacks as well as queues. These assertions include:

- Example 7-3, "PSL FIFO overflow assertion"
- Example 7-5, "PSL FIFO underflow without feed forward"
- Example 7-8, "SystemVerilog legal internal state for valid bits"
- Example 7-8, "SystemVerilog legal internal state for valid bits"

In the following section, we define an additional assertion (unlike assertions for queues) for the case when both a stack push and pop occur, requiring a special sequence order of operations to replace the data as the first entry of the stack.

Example 7-16 ***Verilog* fragment for simple stack**

```
// use a valid vector to indicate entries that are filled (valid)
// as data is pushed in, we shift the stack data and add the new data.
parameter DEPTH    = 16; // 16 entry stack.
parameter WIDTH    = 8;  // 8 bits wide.
parameter STKWIDTH = WIDTH*DEPTH;
reg  [DEPTH-1:0]    valid;
wire [WIDTH-1:0]     data_in;
reg  [STKWIDTH-1:0]  stack_data;
wire [WIDTH-1:0]     top_of_stack = stack_data[WIDTH-1:0];

always @(posedge clk or negedge rst_n) begin
  case ({push, pop, flush})
    3'b100: begin
      valid <= {valid[DEPTH-2:0], 1'b1}; // shift in another entry.
      stack_data <= {stack_data[STKWIDTH-WIDTH-1:0], data_in};
      ...
    3'b001: begin
      valid <= {DEPTH{1'b0}};
      stack_data <= {STKWIDTH{1'b0}};
      ...
    3'b010: begin
      valid <= {1'b0, valid[DEPTH-1:1]}; // pop it out.
      stack_data <= {{WIDTH{1'b0}}, stack_data[STKWIDTH-1:WIDTH]};
      ...
  endcase
```

7.3.1 Assertion

Pattern name. Stack push with pop

Problem. A simultaneous push and pop stack operation may incorrectly replace the top element of the stack.

Motivation. It is easy to overlook the case when push and pop commands occur simultaneously such that the old data is removed from the stack while the new data is added to the top of the stack and valid data depth of the stack does not change.

Solution. In Example 7-17, we check for stack push with pop using a SystemVerilog property. Note that this assertion is actually independent of any specific stack implementation. Hence, not only does it work for our simple stack example, the assertion can be used for a more complex implementation. Note for this example, we are using the new SystemVerilog $stable system function, which returns true if the previous value of its argument is the same as the current value. For additional details, see Appendix C.

Example 7-17 ***SystemVerilog*** **correct push and pop operation of stack**

```
// for a push and a pop, then on the next cycle the top of the
// stack must equal what was previously pushed onto the stack
// and the stack depth is the same
property push_and_pop_good;
  @(posedge clk)
    (push & pop |=> $past(data_in) == top_of_stack && $stable(valid));
endproperty
assert property (push_and_pop_good);
```

7.3.2 Functional coverage

7.3.2.1 Stack depth coverage

Pattern name. Stack depth

Problem. Without tracking coverage measurement for various levels of valid content contained within a stack, we do not know if we have tested boundary conditions for the stack, and we have not tested the architectural performance.

Motivation. We must ensure that boundary conditions for the stack have been adequately tested. If we have not used all elements, we have either over designed the stack (the stack depth is too large), under designed the stack (not large enough, which affects performance), or we have not fully tested our design.

Solution. In Example 7-18., we report the filling for each possible level of the stack to determine the maximum depth achieved during verification. This is useful if we want to measure architectural performance.

functional coverage caution

As previously discussed in the introduction to this chapter, be cautious when specifying this level of functional coverage. We recommend that you only do this if it is critical to determine the architectural performance of a specific design implementation.

Example 7-18 ***PSL*** **functional coverage usage of each element**

```
// see Appendix B for details on the PSL forall construct
cover forall N in {0:16} : {rose(valid[N])};
```

7.3.2.2 Stack flushes

Pattern name. Stack flushes

Problem. If the stack flush control circuit is not designed properly, invalid data could be popped out of the stack.

Motivation. It can be difficult to debug unexpected behavior during simulation. Without testing flush operations, we cannot ensure that flushed content on the stack does not affect subsequent behavior. However, if we are monitoring the occurrence of critical behavior during simulation, we can associate the failure with other events that are occurring about the same time as the failure.

Solution. Example 7-19 reports the occurrence of a flush for various levels of content within the stack.

Example 7-19 ***PSL*** **functional coverage of flushing of each valid entry**

```
cover forall N in {0:7} : {flush & valid[N]};
```

7.4 Caches—direct mapped

Context. Caches are used to reduce the required high memory bandwidth and data latency to a processor's main memory. They are a critical design element required to achieve a desired system performance, particularly in multi-processing systems. Assertions combined with functional coverage ensure that a system design makes full use of its cache elements (that is, achieves the desired performance) and that the integrity of the data is correct within the system (that is, *coherent*).

Example 7-20 is a fragment of a *direct mapped cache* controller. This simple cache example manages data from a specific memory region in segments of eight bytes, which is referred to as a cache *line*. The cache supports both *write* and *read* requests (corresponding to stores and loads). It also supports cache data *invalidate*, as well as updates of the cache from a specific region of memory (referred to as cache *fills*).

A high-level description of cache operations follows. When a processor (requestor) makes a request to the cache, the cache looks up the address in its cache *tag set*. If the requested address matches a *valid* line (referred to as a cache *hit*), the cache returns data from the line to the requesting processor. When there is no hit, the cache makes a request to memory. When this request completes, the cache line is filled with the data from the specific region of memory (a *cache fill*) and returns the data to the requestor. This data is now available in the cache if a similar address (that is, an address within regions of the address for the valid cache line) is requested in the future.

Example 7-20 *Verilog* **fragment for high-level direct mapped cache**

```
parameter ADDRW    = 32;       // number of address bits
parameter NLINES   = 8;        // number of lines
parameter LINE     = 9;        // bytes in a line
parameter logNLINS = 3;        // log based 2 of N lines
parameter TAG      = ADDRW-3;  // address minus 3 index minus 3 offset

reg [8*LINE-1:0]    cache_line[NLINES-1:0];
reg [8*TAG-1:0]     cache_tag;  // tag set
reg [NLINES-1:0]    valid, n_valid, match;
reg [NLINES-1:0]    cache_we;   // write data [line]
reg [logNLINS-1:0] line_sel;   // hit line selector
reg                 hit;        // request hit in cache
wire                cache_fill; // fill line from mem, mark valid
wire                write;      // store request
wire                request;    // request to the cache
wire [ADDRW-1:0]    addr;       // request address
wire                invalidate; // invalidate the cache

always @(*) begin
  case({write, request, cache_fill, invalidate})
     4'b1100,
     4'b0100: begin              // write or read request
  // index is a function returns index from the addr to select a line
         line_sel   = index(addr);               // extract index
  // hit_detect is a function returns vector for tag match
         match      = hit_detect(addr, cache_tag) & valid;
         hit        = |match;                    // compute hit of match
         write_hit = {NLINES{write}} & match; // compute write enable
       ...
     4'b0010: begin                              // fill (from memory)
         n_valid = valid | cache_we;             // setting a new valid
         ...
     4'b0001: begin                              // invalidate
         n_valid = {NLINES{1'b0}};               // clear all valids
         ...
```

7.4.1 Assertions

7.4.1.1 Cache line fill

Pattern name. Cache line fill

Problem. Performance and coherency problems within a cache system occur if a cache *fill* does not result in a *valid* cache line in the absence of an *invalidate* command.

Motivation. We must ensure that the cache is updated from memory when a *cache fill* request occurs. Checking to ensure that a *fill* produces a *valid* line reduces further request latencies.

Solution. Example 7-21 shows an assertion for eight cache lines using a PSL forall construct. A *fill* request for a specific line must produce a *valid* line the next cycle (unless an *invalidate* occurs).

Example 7-21 *PSL* fill (write) implies valid line

```
// for eight cache lines
assert forall N in {0:7} :
  always {cache_fill && cache_we[N] & !invalidate}
    |=> {valid[N]} @(posedge clk);
```

7.4.1.2 No hit on invalid cache lines

Pattern name. No hit on invalid cache lines

Problem. If an invalid line receives a *hit* (that is, address match), and thus returns data, unpredictable behavior can occur, which can be difficult to debug and isolate during verification.

Motivation. Calculation of the cache *hit* result may be distributed among multiple equations where it is easy to overlook required terms for a specific equation. If a the logic calculating a hit is in error, invalid data could be returned from the cache, resulting in an error.

Solution. The cache line selection is determined from an address. However, the valid bit must be included as part of an address match. Example 7-22 shows an assertion that specifies that an inactive valid bit will never result in a hit.

Example 7-22 *SystemVerilog* not valid line implies no hit

```
property no_invalid_hit;
   @(posedge clk) (!valid[line_sel] |-> !hit);
endproperty
assert property (no_invalid_hit);
```

7.4.1.3 Invalidating a cache

Pattern name. Invalidating a cache

Problem. Stale data (that is, data that should have been invalidated) within a cache causes coherency problems in a system.

Motivation. Cache invalidation is used to clear a cache of data that is no longer valid with respect to the main memory store, thus ensuring coherency in a system. Use of stale data causes a system crash, algorithm failure, and difficult to diagnose problem.

Solution. Example 7-23 ensures that once invalidate is issued, all cache lines are *invalid*.

Example 7-23 ***PSL* invalidate implies empty cache**

```
assert always {rose(invalidate)} |=> {!valid};
```

7.4.1.4 Write request update

Pattern name. Write request update

Problem. If an invalid cache line is accessed after a write request update, unpredictable behavior can occur, which can be difficult to debug and isolate during chip-level verification.

Motivation. For writes that *hit* a cache line, depending on the design, the cache may choose to either update the cache data—or may invalidate the cache line (for example, for a write-thru cache design or a hit of a shred line). Reading stale data after an invalidate is a coherency problem.

Solution. Example 7-24 shows a write request producing a *hit* that must cause an update to the cache data—or an *invalidate* to the cache line on the following cycle.

Example 7-24 ***PSL* write hit creates new data or invalid line**

```
// For all eight cache line.
assert forall L in {0:7} :
  always {rose(write_hit[L])} |=>
    {prev(cache_line[L]) != cache_line[L] || !valid[L]};
```

7.4.2 Functional coverage

Functional coverage for a cache serves two purposes. It ensures the logic is operating correctly and that system performance is not being lost (that is, cache performance relies on full use of the cache structure to provide data with minimal latency).

7.4.2.1 Cache line fill

Pattern name. Cache line fill

Problem. Cache lines that are not filled do not contribute to the desired system performance.

Motivation. The performance of a cache system is based on the proper handling of existing data contained within the cache during a request. To ensure good performance, it is critical that we verify that all cache lines have gone from invalid to valid for a new address during a write request.

Solution. Example 7-25 shows a coverage event for a cache *fill* (write) request to a specific cache line (`cache_we[i]`). We have specified functional coverage, which will report each time a requested line is filled during verification.

Example 7-25 *PSL* **functional coverage of each valid line (cache fill)**

```
cover forall N in {0:7} : // for 8 cache lines.
  {cache_fill & cache_we[N]};
```

7.4.2.2 Cache line usage

Pattern name. Cache line usage

Problem. If we do not receive a *hit* on a given cache line, then the cache line does not contribute positively to the cache system performance (that is, it is impossible to get a hit on the line due to a design error or we have not achieved adequate testing of the line).

Motivation. Once a cache line is filled, future requested addresses should be able to *hit* in the cache, ensuring good system performance. Failure to *fill* a line or *hit* a line during verification is cause for further analysis to determine if an error exists in the cache system design.

Solution. Example 7-26 creates functional coverage, which reports the hit for a given cache line. With this and the previous example, we can see if lines are being brought into the cache (that is, was previously filled) and are subsequently reused by the processor.

Example 7-26 *PSL* **functional coverage of each cache line hit**

```
cover forall L in {0:7} : // 8 line cache
  {L == line_sel && valid[L] & hit};
```

7.4.2.3 Write hit to cache line

Pattern name. Write hit to cache line

Problem. If we do not receive a *write* hit on a given cache line, then the cache line has not been adequately tested, or the cache line might have been corrupted.

Motivation. Once a cache line is filled, future addresses should be able to *hit* in the cache. Not receiving a write hit during verification is cause for further analysis to determine if an error exists in the cache system design.

Solution. In Example 7-27, the functional coverage specification ensures sufficient write *updates* have occurred to all cache lines during verification. Sufficient updates, combined with other cache operations to a cache line ensures that the cache data is consistent on future operations. Note how this complements the assertion in Example 7-24, by ensuring that a `write_hit` occurs (that is, we cannot check the assertion in Example 7-24 if the `write_hit` never occurs).

Example 7-27 ***PSL* functional coverage of each cache line written (store - hit)**

```
cover forall L in {0:7} : {!cache_fill & write_hit[L]};
```

7.4.2.4 Cache event timing

Pattern name. Cache event timing

Problem. Without adequate testing of specific timing interaction of cache events, design errors can go undetected.

Motivation. Testing with specific timing relationships between multiple events in a cache can uncover potential corner cases where other unexpected independent events coincide. By reporting the occurrence for specific timing relationships between independent events (for example, a *fill* and a *hit*) you ensure that specific timing of events is not overlooked, which could cause a failure if not exercised.

Solution. Example 7-28 shows functional coverage for a cache *fill* request related to a load request hit of the cache. By checking that these events occurred during verification, you ensure your logic properly handles the specific timing relationships.

Example 7-28 *PSL* functional coverage of timing of events

```
// ensure timing between fill and hit of same line is covered (-2..2)
// request is used in the last two coverage specifications, because we
// can't hit when the fill is not done yet
cover {cache_fill; 1; hit && prev(fill_sel, 2) == line_sel}; // 2cycle
cover {cache_fill; hit && prev(fill_sel) == line_sel};        // 1cycle
cover {cache_fill && hit && fill_sel == line_sel};            // 0cycle
cover {request; cache_fill && prev(line_sel) == fill_sel};    // 1cycle
cover {request; 1; cache_fill &&
                          prev(line_sel, 2) == fill_sel}; // 2cycle
```

7.5 Cache—set associative

Context. A set associative cache, unlike a direct mapped cache, allows for more than one region of memory to be stored in a cache line at a time. Hence, an address can reside in a *set* of locations within the single line (versus a single location for a direct mapped

Example 7-29 *Verilog* fragment for high-level set-associative cache

```
parameter ADDRW     = 32;      // number of address bits
parameter NLINES    = 8;       // number of lines
parameter LINE      = 8;       // bytes in a line
parameter logNLINS  = 3;       // log based 2 of N lines
parameter TAG       = ADDRW-6; // address minus 3 index minus 3 offset
parameter NSETS     = 4;       // number of sets in a line

reg [8*LINE-1:0]    cache_line[NSETS*NLINES-1:0];
reg [NSETS*TAG-1:0]cache_tag[NLINES-1:0];
reg [NSETS-1:0]     valid[NLINES-1:0], n_valid[NLINES-1:0],
                    set_match,  // aset matched
                    set_valid,  // valid bits of selected line
                    write_set;  // write a set
reg [NLINES-1:0]    cache_we;   // write a line
reg [logNLINS-1:0] line_sel;    // hit line selector
reg                 hit;        // request hit in cache
wire                cache_fill; // fill line from mem, mark valid
reg [logNLINS-1:0] fill_sel;    // line to fill
wire                write;      // store request
wire                request;    // request to the cache
wire [ADDRW-1:0]    addr;       // request address
wire                invalidate; // invalidate the cache

always @(*) begin
  for(i=0;i<NLINES; i=i+1) n_valid[i] = valid[i];
  case({write, request, cache_fill, invalidate})
     4'b1100,
     4'b0100: begin             // write or read request
  // index is a function returns index from addr for line selection
          line_sel  = index(addr);         // extract index
          set_valid = valid[line_sel];
  // set_hit_detect is a function returns vector for a tag match
          set_match = set_hit_detect(addr, cache_tag[line_sel])
                      & set_valid;
          hit       = |set_match;          // compute hit of match
          write_hit = {WIDTH{write}} & set_match; // compute enable
        ...
     4'b0010: begin                        // fill (from memory)
          n_valid   = valid | cache_we;    // setting a new valid
          ...
     4'b0001: begin                        // invalidate all valid bits
          for(i=0;i<NLINES;i=i+1) n_valid[i] = {NSETS{1'b0}};
          ...
```

cache). A line now contains multiple address sets, where each set contains data for specific addresses that map into a single line. Thus a set associative cache can store multiple sequences of code that would otherwise conflict for the cache space of a direct mapped cache. This allows for a specific software routine to complete its execution while still referencing data in the cache, even though some other concurrent executing code may be accessing data from a different address space loaded into the same cache line.

The overall operation for a set associative cache is the same as the direct mapped cache described in Section 7.4 "Caches—direct mapped". A requestor provides an address to the cache that is compared to the tags in the cache, which represents the data from a specific region of memory currently loaded into the cache line. In the set associative cache, multiple tags are compared to the specified address, and only one set from the cache line can match (and *hit*). When a hit does occur, the data from that particular set is returned. When a miss occurs, a request to a region of memory is made and the data from memory is loaded into a specific set (chosen by the cache line replacement policy).

A set associative cache, in general, has challenges similar to a direct mapped cache. These problems include efficient usage of the entire cache and correctness of *fills*, *hits* and *invalidates* of the cache. (We demonstrated assertion and coverage techniques for these problems in the previous section.) In this section, we focus on specifying assertions and coverage specifically related to set associative caches.

7.5.1 Assertion

Pattern name. Match a single set within a line

Problem. If a single region of memory are mapped into multiple sets within the same cache line (that is, a request address matches multiple tags within the line), coherency in the system cannot be achieved.

Motivation. A line in a set associative cache contains several addresses. These addresses must be unique for correctness (and maximum performance). Hence, there must not be more than one (if any) matches for a given requested address to a tag within the cache line.

Solution. Example 7-30 specifies an assertion for the *set match signal* (that is, `set_match`) to ensure that it has matched the requested address to no more than a single tag from a set within the line.

Example 7-30 ***SystemVerilog*** **at most only a single set match allowed**

```
property onehot_match;
  @(posedge clk) ($countones(set_match[3:0])<=1);
endproperty
assert property (onehot_match);
```

7.5.2 Functional coverage

7.5.2.1 Cache set fill

Pattern name. Cache set fill

Problem. Cache sets that are not filled do not contribute to the desired system performance.

Motivation. The performance of a cache system is based on the proper handling of existing data within the cache during a request. To ensure good performance, it is critical to verify that all cache lines have gone from invalid to valid for a new address during a write request.

Solution. Example 7-31 shows a coverage event for a cache *fill* (write) request to a specific set within a cache line (`write_set[S]`). We have specified functional coverage that will report each time a requested set is filled during verification.

Example 7-31 ***PSL*** **functional coverage for each set written**

```
// for all lines--for all sets
cover forall L in {0:7} :
        forall S in {0:3} : cache_we[L] & write_set[S] & cache_fill};
```

7.5.2.2 Usable cache set

Pattern name. Usable cache set

Problem. If we do not receive a *hit* on a given set within a cache line, then the set does not contribute positively to the cache system performance (that is, it is impossible to get a hit on the set due to a design error or we have not achieved adequate testing of the set).

Motivation. Once a set within cache line is filled, future requested addresses should be able to *hit* in the cache, ensuring good system performance. Failure to *fill* a set or *hit* a set during verification is cause for further analysis to determine if a error exist in the cache system design.

Solution. Example 7-32 creates functional coverage, which reports a hit for a given set within a cache line.

Example 7-32 ***PSL* functional coverage for each set hit**

```
// for all lines--for all sets
cover forall L in {0:7} :
        forall S in {0:3} : {L == line_sel & set_match[S] & hit};
```

7.5.2.3 Write hit to set

Pattern name. Write hit to set

Problem. If we do not receive a *write* hit to a given set within a cache line, then the set has not been adequately tested, or the cache set might have been corrupted.

Motivation. Once a set within a cache line is filled, future addresses should be able to *hit* in the cache. Not receiving a write hit during verification is cause for further analysis to determine if a error exist in the cache system design.

Solution. These events in Example 7-33 ensure that sufficient use of the set within a cache lines (updated when already *valid*) is done to ensure that the cache works more than once and is robust.

Example 7-33 ***PSL* functional coverage for each valid set (of a line) written**

```
// for all lines--for all sets
cover forall L in {0:7} :
  forall S in {0:3} :
    {!cache_fill & cache_we[L] & write_set[S] & set_valid[S]};
```

7.5.2.4 Reuse of existing set

Pattern name. Reuse of existing set

Problem. To ensure good performance, each set within a cache line must either go from *invalid* to *valid* for the new requested address—or from a different *valid* address currently contained within the cache to the new requested address not contained within the cache.

Motivation. Ensuring the cache line's *set replacement policy* is operating correctly is another performance aspect that is difficult to diagnose. By adding functional coverage, we can ensure that the this functionality is being tested.

Solution. In Example 7-34, the functional coverage specification ensures sufficient write *updates* have occurred to all cache lines during verification. Sufficient updates, combined with other cache operations, to a cache line ensures that the cache data is consistent on future operations.

Example 7-34 ***PSL* functional coverage for each valid set (of a line) filled**

```
cover forall L in {0:7} :
  forall S in {0:3} :
    {cache_fill & cache_we[L] & write_set[S] & set_valid[S]};
```

7.6 FSM

Context. Finite state machines offer many situations that are ideal for specifying correct operation using assertions and functional coverage. Some of the functional coverage examples shown in this section may be more effectively measured using commercial coverage tools. However, complex interactions between state machincs is generally difficult to measure with commercial tools and are better suited to functional coverage specification.

Figure 7-2 **FSM example used in the following sections**

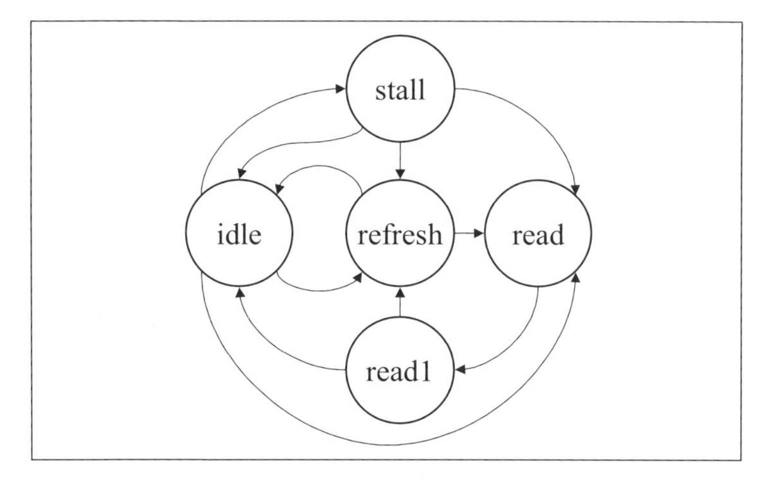

7.6.1 Assertions

7.6.1.1 Legal FSM states

Pattern name. Legal FSM state

Problem. If an FSM enters an illegal state, unpredictable behavior can occur, which can be difficult to debug and isolate at the chip-level during verification.

Motivation. Often, the designer focuses on only the legal states of an FSM during the design process—or makes assumptions that illegal states are not possible. Without trapping situations that cause an FSM to enter an illegal state, a significant amount of debug effort is generally required to isolate (and map) a chip-level failure down to the offending FSM.

Solution. In Example 7-35, we apply a version of the *valid opcode pattern* to specify valid states for the FSM illustrated in Figure 7-2.

Example 7-35 ***SystemVerilog* check for only legal states**

```
property legal_states;
  @(posedge clk) disable iff !rst_n
     ($inset(state, 'IDLE, 'READ, 'READ1, 'REFRESH, 'STALL));
endproperty
assert property (legal_states);
```

7.6.1.2 Inputs for specific states

Pattern name. Inputs for specific states

Problem. An illegal input value for a specific FSM state can cause unpredictable behavior, which can be difficult to debug and isolate at the chip-level during verification.

Motivation. Generally, FSMs only allow certain combinations of input values when the state machine is in a specific state. For example, assume the FSM shown in Figure 7-2 does not allow either a new `req` or `abort` to occur when the FSM is in a `READ` or `READ1` state. If a new `req` or `abort` occurs, then the FSM might enter into an illegal state—or make an unexpected transition to a different legal state.

Solution. Example 7-36 demonstrates how to code an assertion that ensures a `req` or `abort` is not received as an input while the state FSM is in either a `READ` or `READ1` state.

Example 7-36 ***SystemVerilog* check for specific inputs only in specific states**

```
property illegal_input; // wrong time for input
  @(posedge clk) not ($inset(state, 'READ, 'READ1) & (req | abort));
endproperty
assert property (illegal_input);
```

7.6.1.3 State transition performance limits

Pattern name. State transition performance limits

Problem. Failure to satisfy a state transition timing requirement may indicate an error situation (for example, a deadlock).

Motivation. Often, interfaces are designed under the assumption that they will meet a specific performance requirement. For example, a protocol controller might require that an acknowledge is received within a fixed number of specified clocks after a request—or a memory controller is expected to return to an idle state within a specified limited range of cycles after the start of a refresh cycle. Failure to satisfy the timing requirements often indicates a forward progress or deadlock situation.

Solution. In Example 7-37, we demonstrate how to specify a performance limit for a state transition.

Example 7-37 ***SystemVerilog* check for operation to complete in fixed time periods**

```
property trans_toidle;
  @(posedge clk) ($rose(state=='REFRESH) |-> ##[5:15] state=='IDLE);
endproperty
assert property (trans_toidle);
```

7.6.2 Functional coverage

7.6.2.1 FSM state coverage

Pattern name. FSM state coverage.

Problem. Without visiting all states of an FSM during the course of verification, potential bugs may go undetected.

Motivation. It is important during the course of verification to visit all legal states within a given FSM. Otherwise major functionality could go untested prior to tapeout.

Solution. In Example 7-38 we demonstrate how to write a functional coverage specification, which is used to report the visits to legal states within a given FSM. Note for simple FSMs, that follow a well defined coding style, specifying state coverage is unnecessary if you use a commercial tool that provides this form of coverage. However, complicated state machines or unconventional coding styles typically require explicit functional coverage specification.

Example 7-38 ***PSL* functional coverage each state entered**

```
cover {state == 'IDLE};
cover {state == 'READ};
cover {state == 'READ1};
cover {state == 'REFRESH};
cover {state == 'STALL};
```

7.6.2.2 State sequence coverage

Pattern name. State sequence coverage

Problem. Without specifying expected state sequences (for example, transactions), major functionality within the design could go unverified.

Motivation. The occurrence of specific state sequences or transactions may be difficult to identify in a simulation run. By specifying expected sequences, we can be assured that sufficient verification has occurred and identify functionality within the design that has not been adequately tested.

Solution. Example 7-39 reports the occurrence for any transition from a READ state to a REFRESH state three cycles later. Note that this would potentially be equivalent to going through the sequence: READ -> READ1 -> IDLE -> REFRESH, shown in Figure 7-2.

Example 7-39 ***PSL* ensure proper state sequence**

```
cover {(state == 'READ); [*2]; state == 'REFRESH};
```

Example 7-40 demonstrates how to enumerate the functional coverage specification for specific state transitions. These transitions are related to Figure 7-2.

Example 7-40 ***PSL* functional coverage for each state transition**

```
cover {state == 'IDLE ; state == 'READ};
cover {state == 'READ1 ; state == 'STALL}; // and so on.
```

7.7 Counters

Context. Counters are generally thought of as minor structures within a design. However, counters often have well defined properties (for example, no overflow or no underflow) that, if violated, can cause unpredictable behavior. Assertions enable us to check for the correctness for these properties during verification. In addition, often a specific counter value relates to the occurrence of a specific event within a design. Hence, it is often desirable to specify functional coverage on certain counter values.

The following code fragment shows an up-down counter with a parallel load option.

Example 7-41 ***Verilog*** **fragment up-down counter with parallel load**

```
parameter max=5'd26;
reg [4:0] count;
always @(posedge clk or negedge rst_n) begin
  if (!rst_n)
    count <= 5'b0;
  else
    casez ({inc, dec, load})
      3'b??1: count <= count_in; // load takes priority.
      3'b100: count <= count + 1'b1;
      3'b010: count <= count - 1'b1;
      //0, 6 - The default case keeps the same counter value.
    endcase
end
```

7.7.1 Assertions

7.7.1.1 Maximum counter value

Pattern name. Maximum counter value

Problem. A counter that exceeds its maximum value (for example, wraps around to a lower value), can cause lost data if used as a data pointer—or, in general, cause unpredictable behavior, which can be difficult to debug or isolate at the chip-level during verification.

Motivation. Note in Example 7-41 (above) that no maximum value for `count` is enforced. Hence, the counter will wrap (that is, roll over from its maximum value) if incremented on its maximum value. This might be the desired behavior of the counter (for example, if we have implemented a counter for a circular queue). However, if this is not the desired behavior, then we should add an assertion to ensure that the circuitry preventing the illegal increment is functioning correctly.

Solution. Example 7-42 demonstrates the coding of an assertion in SystemVerilog that ensures that a counter does not exceed a specified maximum value. In addition, this example demonstrates how to code an assertion to check for an overflow condition.

Example 7-42 *SystemVerilog* counter limits

```
property exceed_max;
  @(posedge clk) not (count > max);
endproperty

property no_overflow;
  @(posedge clk) not (count == 0 && $past(inc & !load, 1));
endproperty

assert property (exceed_max);
assert property (no_overflow);
```

7.7.1.2 Counter underflow

Pattern name. Counter underflow

Problem. A counter that underflows its minimum value (for example, wraps around to a higher value), can cause lost data if used as a data pointer—or, in general, cause unpredictable behavior, which can be difficult to debug or isolate at the chip-level during verification.

Motivation. Note in Example 7-41 (above) that no minimum value for count is enforced. This might be the desired behavior of the counter (for example, if we have implemented a counter for a circular queue). However, if this is not the desired behavior, then we should add an assertion to ensure that the circuitry preventing illegal decrement is functioning correctly.

Solution. Example 7-43 demonstrates the coding of an assertion in PSL that ensures that a counter does not underflow.

Example 7-43 *PSL* no wraparound (underflow)

```
assert never (count == 0 & dec & !load);
```

7.7.2 Functional coverage

In some designs, specific values within a counter are associated with critical events. Hence, functional coverage for specific values helps identify the occurrence of these critical events during verification.

7.7.2.1 Counter limits

Pattern name. Counter limits

Problem. If the boundary conditions for a counter are not adequately verified, unpredictable behavior can occur, which can be difficult to debug or isolate at the chip-level during verification.

Motivation. In many designs, counters are used as pointers into other structures, such as memory addressing, FIFO, and stack pointers. Testing the boundary conditions for these counters is critical during verification to flush out corner case errors.

Solution. In Example 7-44 we demonstrate how to specify functional coverage for the boundary conditions on a counter.

Example 7-44 ***PSL* functional coverage for boundary condition**

```
cover {count==max-2; count==max-1};
cover {count==1; count==0};
```

7.7.2.2 Counter control

Pattern name. Counter control

Problem. If all control combinations for a counter are not exercised, then a corner case error might be missed during verification.

Motivation. For non-trivial counters, or counters associated with complex control logic, we might wish to know what combinations of controls for the counter have been exercised. Hence, the counter control could give us a level of confidence in testing the complex circuit.

Solution. Example 7-45 demonstrate how to specify functional coverage on the controls related to our counter example. Note that for simple circuits, this would probably be too much functional coverage data, as we pointed out in the introduction to this chapter. Yet for key portions of a design, we might want to determine exactly what combination of events are occurring during verification.

Example 7-45 ***PSL* functional coverage for increment/decrement conditions**

```
cover forall N in {0:7} : {{inc, dec, load} == N};
```

7.8 Multiplexers

Context. Multiplexers are key structures used to conditionally control the flow of data through a design. In this section, we discuss three common types of multiplexers, whose implementation differs based on the design of the data selection circuit (for example, *encoded*, *decoded*, or *prioritized*).

7.8.1 Encoded multiplexer

In Example 7-46, we demonstrate the RTL for an encoded multiplexer. We use this example during our following discussion.

Example 7-46 ***Verilog* fragment for encoded multiplexer code**

```
always @(select or data0 or data1 or data2 or data3)
  case(select)
    2'd0: outdata = data0;
    2'd1: outdata = data1;
    2'd2: outdata = data2;
    2'd3: outdata = data3;
  endcase
```

7.8.1.1 Functional coverage

Pattern name. Encoded selector coverage

Problem. Without exercising all critical control paths within a design during verification, corner cases can go undetected.

add functional coverage to RTL only when it makes sense

Motivation. For critical control paths, it is necessary to ensure all possible paths are exercised. Code coverage, is ideal for many of these lower-lever structures, like encoded multiplexers. However, if a particular multiplexer structure is deemed sufficiently critical (particularly related to a complex design with corner case concerns), functional coverage for a particular encoded multiplexer could be specified.

Solution. In Example 7-47, we demonstrate how to specify functional coverage for all encoded multiplexer `select` lines. Hence, we can easily determine during verification if a particular critical path has not been exercised.

Example 7-47 ***PSL* functional coverage for a four-to-one encoded mux**

```
cover forall N in {0:3} : {select == N};
```

7.8.2 Decoded (one-hot) multiplexer

Context. Decoded multiplexers uses a single select bit (from a multi-bit value) as an enable to transfer the selected input data to an output. Example 7-48 demonstrates the RTL for a simple decoded multiplexer, which is used in the following related discussion on assertions and functional coverage.

Example 7-48 ***Verilog* fragment for decoded multiplexer example**

```
always @(select or data0 or data1 or data2 or data3)
  case(1'b1)
    select[0]: outdata = data0;
    select[1]: outdata = data1;
    select[2]: outdata = data2;
    select[3]: outdata = data3;
  endcase
```

7.8.2.1 Assertion

Pattern name. Decoded valid selector value

Problem. If more than one selector value is active for a one-hot multiplexer, then potentially the data could be corrupted—or the multiplexer circuit could be damaged (for example, a tri-state implementation).

Motivation. A decoded multiplexer are often used in a design to achieve a required timing performance, in contrast to a slower encoded multiplexer. However, for proper operation, it is critical that only a single select line is enabled at a time for the decoded multiplexer. Otherwise, the multiplexed data might be corrupted—or the multiplexer circuit is might be damaged.

Solution. In Example 7-49, we demonstrate how to specify an assertion to validate the decoded multiplexer's one-hot select line requirement. For this example, we use the SystemVerilog $onehot system function.

Example 7-49 *SystemVerilog* assertion to validate one-hot selector

```
assert property ($onehot(select));
```

7.8.2.2 Functional Coverage

Pattern name. Decoded input selector values

Problem. Without exercising all critical control paths within a design during verification, corner cases can go undetected.

add functional coverage to RTL only when it makes sense

Motivation. For critical control paths, it is necessary to ensure all possible paths are exercised. Code coverage, is ideal for many of these lower-lever structures, like decoded multiplexers. However, if a particular multiplexer structure is deemed sufficiently critical (particularly related to a complex design with corner case concerns), functional coverage for a particular decoded multiplexer could be specified.

Solution. In Example 7-47, we demonstrate how to specify functional coverage for all decoded multiplexer `select` lines. Hence, we can easily determine during verification if a particular critical path has not been exercised.

Example 7-50 *PSL* functional coverage for each input selected

```
cover {select[2'b00]};
cover {select[2'b01]};
cover {select[2'b10]};
cover {select[2'b11]};
```

7.8.3 Priority multiplexer

Context. In many designs, certain events and interrupts are categorized such that higher-priority events execute prior to lower-priority events. In these types of designs, priority multiplexers are useful for routing the appropriate event in a prioritized order. In Example 7-51, we demonstrate a simple priority multiplexer, which is used in the following related discussion on assertions and functional coverage.

Example 7-51 *Verilog* **fragment for priority multiplexer**

```
always @(select or data0 or data1 or data2 or data3)
  casez(select) // synopsys full_case parallel_case
    4'b???1: outdata = data0;
    4'b??10: outdata = data1;
    4'b?100: outdata = data2;
    4'b1000: outdata = data3;
    default: assert property (@(posedge 'TOP.clk) (1'b0));
          // Not legal select value.
  endcase
```

7.8.3.1 Assertion

Pattern name. Priority legal selection values

Problem. If a zero multiplexer selector value is driven from some controlling logic, then potentially the data could be corrupted, or unpredictable behavior, which can be difficult to debug or isolate at the chip-level during verification.

Motivation. A priority multiplexers are useful for routing the appropriate event in a prioritized order. However, for proper operation, it is critical that at least one of the select bits is enabled.

Solution. In Example 7-51, we demonstrated the RTL for a priority multiplexer, with a procedural assertion. In Example 7-52, we demonstrate a declarative form for this assertion.

Example 7-52 *SystemVerilog* **ensure at least one select is active**

```
property atleastone;
  @(posedge clk) not (select == 4'b0);
endproperty

assert property (atleastone);
```

7.8.3.2 Functional Coverage

Pattern name. Priority input select values

Problem. Without exercising all critical control paths within a design during verification, corner cases can go undetected.

add functional coverage to RTL only when it makes sense

Motivation. For critical control paths, it is necessary to ensure all possible paths are exercised. Code coverage, is ideal for many of these lower-lever structures, like priority multiplexers. However, if a particular multiplexer structure is deemed sufficiently critical (particularly related to a complex design with corner case concerns), functional coverage for a particular priority multiplexer could be specified.

Solution. In Example 7-53, we demonstrate how to specify functional coverage for all priority multiplexer `select` lines. Hence, we can easily determine during verification if a particular critical path has not been exercised.

Example 7-53 ***PSL*** **functional coverage for all combinations of selects**

```
// zero is illegal for the select
cover forall N in {1:15} : {select == N};
```

7.8.4 Complex multiplexer

Context. Multiplexers in many of today's designs are usually not as simple as the previous examples. Often, the design requires special needs to handle complex situations when routing control or data. Multiplexing structures, such as a Verilog case statement, are typical used to describe these complex routing needs. In Example 7-54, we demonstrate a complex multiplexer, which uses a Verilog casez construct.

Example 7-54 ***Verilog-2001*** **fragment for complex multiplexer**

```
always @(*) begin
  ...
  legal_state_size: assert property
    ( @(posedge clk) disable iff (reset_n)
      ($onehot0(state_size1[2:0])))
    else $error("illegal state_size1 value %0b.", state_size1[2:0]);

  casez({size_sel, state_size1[2:0]}) // synopsys parallel_case full_case
    4'b1_??1: next_size = case_a_route;
    4'b0_??1: next_size = case_b_route;
    4'b?_1??: next_size = case_c_route;
    4'b?_?1?: next_size = case_d_route;
    4'b0_000: next_size = case_e_route;
  endcase
end
```

7.8.4.1 Assertion

Pattern name. Complex multiplexer legal selection

Problem. If design assumptions are invalid, control logic can experience unexpected behavior during verification.

Motivation. Use of casez (or casex) allows specification of don't care values on signals. This is done to reduce the selection logic. Often, assumptions are made about legal and select possibilities during the initial design. However, these assumptions need to be validated. Furthermore, as the design is modified in the future, the original assumptions should be re-verified.

Solution. The assertion shown in Example 7-54 specifies that no more than one bit of the `state_size1` variable is set to one. This example uses the SystemVerilog $onehot0 system task.

7.9 Encoder

Context. An encoder generally converts a sparse or uncompressed bit representation into a more dense or compressed format, which is useful when transferring control information between multiple components in a design. For example, a four-bit one-hot state machine encoding can be converted into a more compressed two-bit decimal encoding for communication use, as shown in Example 7-55.

Example 7-55 ***Verilog*** **fragment for a two-bit encoder**

```
always @(control_state)
 // Convert the decoded input to encoded form.
 case(1'b1)
   control_state[0]: encode2 = 2'd0;
   control_state[1]: encode2 = 2'd1;
   control_state[2]: encode2 = 2'd2;
   control_state[3]: encode2 = 2'd3;
 endcase
```

7.9.1 Assertion

Pattern name. Encoder legal input value

Problem. If the value being encoded is in error, then the component receiving the encoded value can exhibit unpredictable behavior, which can be difficult to debug or isolate at the chip-level during verification.

Motivation. Encoder logic within the RTL design often makes assumptions that the input data is correct. If there is an unexpected error with the input encoding, then this can result in unpredictable behavior.

Solution. In Example 7-55, our RTL encoder assumes that its `control_state` variable has a one-hot value. Hence, in Example 7-56 we demonstrate how to write a SystemVerilog assertion to validate this requirement.

Example 7-56 ***SystemVerilog* validate encoder legal input values**

```
valid_encoder_input:  assert property
    (@(posedge clk) $onehot(control_state));
```

7.9.2 Functional coverage

Pattern name. Encoder input values

Problem. Without exercising all encoder input values within a design during verification, corner cases can go undetected.

add functional coverage to RTL only when it makes sense

Motivation. For complex encoders (unlike our simple encoder), it might be necessary to identify which values have not been exercised. For simpler encoders, this should be avoided, since code coverage tools generally provide enough feedback on the quality of their testing. However, if a particular encoder structure is deemed sufficiently critical or complex, then functional coverage for these special cases could be specified.

Solution. In Example 7-47, we demonstrate how to specify functional coverage for all encoded inputs. Hence, we can easily determine during verification if a particular critical path has not been exercised.

Example 7-57 ***PSL* functional coverage for each value received**

```
cover forall N in {0:3} : {encode2 == N};
```

7.10 Priority encoder

Context. A priority encoder converts a prioritized sparse or uncompressed bit representation into a more dense or compressed format, which is useful when transferring control information between multiple components in a design. For example, a combination of interrupts could be grouped into a prioritized order, and then encoded into a request, which is then transferred to other components. Note that using the priority encoder in this fashion is actually a simple form of an arbiter.

In Example 7-55, we demonstrated a priority encoder, which is used in the following related discussion on functional coverage.

Example 7-58 *Verilog* fragment for a priority two-bit encoder

```
always @(int1 or int2 or int3 or int4)
  casez({int1, int2, int3, int4}) // sysnopsys full_case
    4'b???1: int_req2 = 2'd0;
    4'b??10: int_req2 = 2'd1;
    4'b?100: int_req2 = 2'd2;
    4'b1000: int_req2 = 2'd3;
  endcase
```

7.10.1 Functional coverage

Pattern name. Priority encoder input values

Problem. Without exercising all priority encoder input values within a design during verification, corner cases can go undetected.

add functional coverage to RTL only when it makes sense

Motivation. For complex priority encoder (unlike our simple priority encoder), it might be necessary to identify which values have not been exercised. For simpler encoders, this should be avoided, since code coverage tools generally provide enough feedback on the quality of their testing. However, if a particular priority encoder structure is deemed sufficiently critical or complex, then functional coverage for these special cases could be specified.

Solution. In Example 7-59, we demonstrate how to specify functional coverage for all priority encoded inputs. Hence, we can easily determine during verification if a particular critical path has not been exercised.

Example 7-59 *PSL* functional coverage for each select value received

```
cover forall N in {1:15} : {{int1,int2,int3,int4} == N};
```

7.11 Simple single request protocol

Context. A simple single request protocol allows synchronization between multiple components. In the following sections, we demonstrate how to specify assertions and functional coverage related to simple single request protocols.

7.11.1 Assertions

7.11.1.1 Simple handshake

Pattern name. Simple handshake

Problem. A request-acknowledge handshake protocol violation can cause data corruption or data lost, which can be difficult to debug or isolate at the chip-level during verification.

Motivation. The controlling protocol circuits a component communicating other system components can be complex, and error prone. Problems typically encountered include lack of a returned acknowledgement within an expected time limit, dropped acknowledgements during communication, or unexpected multiple request. These problems can result in data corruption or data lost, which can be difficult to debug or isolate at the chip-level during verification.

Solution. We use the pipelined_reqack module, previously defined in Example 6-44 on page 201, to monitor the correct pipeline handshake protocol behavior. For our example, we only use the request (`req`) and acknowledge (`ack`) ports. We specify a maximum latency of 100 cycles, and the maximum number of possible outstanding request as 1 (that is, the depth of the pipelining). This monitor will report violations for the following conditions:

- multiple request without a matching acknowledge
- multiple acknowledges for a single request
- acknowledge not received in the specified maximum limit.

Example 7-60 ***Verilog*** **module to validate req-to-done transaction**

```
// See Example 6-44 on  page 201 for details of pipelined_reqack
// module
pipelined_reqack
   sendReadReq(.req(req),       // The handshakes, request
               .ack(done),       // and done (completion)
               .req_datain(1'b1), .dataout(1'b1), // not used
              .clk(clk),
              .latency(100),  // Maximum latency for return of done.
               .pipedepth(1));  // Only 1 request at a time.
```

7.11.1.2 Legal interface instruction

Pattern name. Legal interface instruction

Problem. An illegal instructions transferred with a protocol request can result in unexpected behavior, which can be difficult to debug or isolate at the chip-level during verification.

Motivation. Some protocols transfers information (such as an instruction) during a protocol request. With complex FIFO designs, it is possible that an illegal instructions is transferred with a request.

Solution. In Example 7-61, we ensure the instruction (`cmd`) in a protocol request is contained within a legal set of instructions. We use `$insetz()`, instead of `$inset()`, to allow for the legal instructions to specify don't care bits in their codes. This can shorten the list of legal values.

Example 7-61 ***SystemVerilog* check for legal information sent**

```
property legal_cmds;
  @(posedge clk) disable iff (reset_n)
    (req |-> $insetz(cmd, 'READ, 'WRITE, 'INTA, 'EIEIO));
endproperty
assert property (legal_cmds);
```

7.11.2 Functional Coverage

Pattern name. Timing between single request

Problem. If the starting of a new transaction can occur within a specified a range, then if the various range of possibilities are not explored, corner cases will not be flushed out.

Motivation. Often, corner case bugs are related to complex combination of events, which are unexpected. The completion of one transaction may affect the next transaction after a specific period. Hence, it is important to cover various ranges of possibilities during verification to ensure a particular period is not overlooked.

Solution. In Example 7-62, we demonstrate how to specify functional coverage for the case where a new request can occur anywhere between zero and twenty cycles after the completion of a previous cycle.

Example 7-62 ***PSL* functional coverage for done-to-req timing (end to begin)**

```
cover forall N in {0:20} : {done; [*N]; req};
```

7.12 In-order multiple request protocol

Context. In-order multiple request protocols allow for greater throughput on an interface by allowing a limited number of transactions to start prior to the completion of a previous transaction. Yet, matching the appropriate transaction completion event with an appropriate request is problematic. In Section 6.7.4, "Pipelined protocol pattern" on page 199 we demonstrated how to code assertions for an in-order multiple request protocol. In this section we demonstrate how to specify functional coverage.

7.12.1 Functional Coverage

7.12.1.1 Multiple request transaction timing

Problem. If the ending of a specific transaction can occur within a specified a range, then if the various range of possibilities are not explored, corner cases will not be flushed out—particularly related to multiple in-order overlapping transactions.

Motivation. Often, corner case bugs are related to complex combination of events, which are unexpected. The completion of one transaction may affect the next transaction after a specific period. Hence, it is important to cover various ranges and overlapping of possibilities during verification to ensure a particular period is not overlooked.

Example 7-63 ***PSL*** **functional coverage for req-to-done timing**

```
reg [3:0] req_cnt, ack_cnt;
initial {req_cnt, ack_cnt} = 8'b0;
always @(posedge clk) begin
  // Increment counter each time a req or ack occurs.
  if (req) req_cnt <= req_cnt + 1;
  if (ack) ack_cnt <= ack_cnt + 1;
end

/* PSL
cover forall C in {0:15} :
  forall T in {0:9} :
    {req && req_cnt == C; [*T]; ack && ack_cnt == C};
*/
```

Solution. In Example 7-63, we specify functional coverage, which enables us to identify the range of occurrences of an acknowledge response for up to sixteen in-order overlapping transactions.

7.12.1.2 Timing of outstanding request

Pattern name. Timing of outstanding request

Problem. For an in-order multiple request protocol, it is critical that multiple outstanding request occur in varying timing relationships to each other during verification to detect potential design errors.

Motivation. Uncovering errors within a design often depends on the generation of stimulus with unexpected alignment of complex events. By reporting the timing relationship from one request to another for an in-order multiple request protocol, we can determine which cycle relationships have not been explored—thus enabling us to tune our verification environment to improve coverage.

Solution. In Example 7-64, we demonstrate how to specify functional coverage used to report various timing occurrences (between zero and ten cycles) for multiple outstanding request. We use a counter as part of our functional coverage model to determine when an outstanding request is present.

Example 7-64 ***PSL* functional coverage for outstanding request timing**

```
reg [3:0] out_cnt;
always @(posedge clk) begin // Track outstanding req's with counter.
  if (reset_n) begin
    out_cnt <= 0;
  end
  else begin
    if (req & !done)  out_cnt <= out_cnt + 1;
    if (done & ! reg) out_cnt <= out_cnt - 1;
  end
end

/* PSL
default clock = (posedge clk);
cover forall N in {0:9} :
   {req; (!done || out_cnt>1) [*N]; req};
*/
```

7.12.1.3 Maximum outstanding in-order transactions

Pattern name. Maximum outstanding in-order transactions

Problem. If we do not check the boundary condition, when the protocol's maximum supported outstanding request has occur, then a corner case could go undetected,.

Motivation. Determining if we utilize the full depth of the queues within a protocol's interface is important—not only for verification, but for performance reason—to determine if the queues are too deep or not deep enough.

Solution. In Example 7-65, specify functional coverage that will report each time the situation when maximum number of supported request has occurred during verification. Our goal is to ensure that full system (the limit of outstanding transactions) has occurred to ensure all operations are functioning as expected.

Example 7-65 ***PSL* functional coverage for max outstanding req's**

```
reg [3:0] req_cnt;
always @(posedge clk) begin
  if (reset_n) begin
    out_cnt <= 0;
  end
  else begin
// Increment counter each time event is seen.
    if (req & !ack)        req_cnt <= req_cnt + 1;
    if (ack & !req)        req_cnt <= req_cnt - 1;
  end
end

// PSL cover {rose(req_cnt) == MAX_OUTSTANDING};
```

7.12.1.4 Timing between multiple requests

Problem. If the starting of a new transaction can occur within a specified a range of cycles, then if the various range are not explored, the a corner case could go undetected.

Motivation. Often, corner case bugs are related to complex combination of events, which are unexpected. The completion of one transaction may affect the next transaction after a specific period. Hence, it is important to cover various ranges of possibilities during verification to ensure a particular period is not overlooked.

Solution. In Example 7-67, we demonstrate how to specify functional coverage for the case where a new request can occur anywhere between one and ten cycles after the completion of a previous cycle.

Example 7-66 ***PSL* functional coverage for done-to-next-req timing**

```
cover forall N in {0:9} :
  {done ; [*N]; req};
```

7.13 Out-of-order request protocol

Context. To improve throughput, many protocols allow multiple transactions to complete in an out-of-order fashion. For example, a tag (that is, unique ID) is often associated with a transaction's initial request, while another tag is associated with the transaction's ending response. To ensure that no data is lost while processing the transaction, it is necessary to validate that for any given tag associated with an initial request, there eventually exists an ending tag with the same ID value. In our pattern discussion in Section 6.7.3, "Tagged transaction pattern" on page 196, we demonstrate how to write assertions for out-of-order protocols. We now demonstrate how to specify functional coverage for these protocols.

7.13.1 Functional coverage

7.13.1.1 Maximum outstanding out-of-order transactions

Pattern name. Maximum outstanding out-of-order transactions

Problem. If all tags have not been used (that is, observed) during verification, then either the out-of-order interface has not been thoroughly test, or the queue depths within the interface were over designed, and the maximum outstanding limit cannot be achieved

Motivation. Determining if we utilize the full depth of the queues within a protocol's interface is import to determine—not only for verification, but for performance reason to determine if the queues or too deep or not deep enough.

Solution. In Example 7-67, we demonstrate how to specify functional coverage to report the occurrence of any of eight possible tags observed during an out-of-order transaction. Many commercial functional coverage tools limit the number of times it reports a specific occurrence of a functional coverage point in simulation. If your tool does not offer this feature, the functional coverage specification could be modified in such a way where you could add in your own limit mechanism.

Example 7-67 ***PSL*** **functional coverage for each tag used**

```
cover forall T in {0:7} : {req && req_tag == T};
```

7.13.1.2 Timing between multiple requests

Problem. If the completion of a transaction can occur within a specified a range, then if the various range of possibilities are not explored, corner cases will not be flushed out.

Motivation. Often, corner case bugs are related to complex combination of events, which are unexpected. The completion of one transaction may affect the next transaction after a specific period. Hence, it is important to cover various ranges of possibilities during verification to ensure a particular period is not overlooked.

Solution. Example 7-68 we demonstrate how to specify functional coverage for the case where an out-of-order acknowledge can occur anywhere between one and ten cycles after the its initial request.

Example 7-68 ***PSL*** **functional coverage for req-to-done timing**

```
default clock = (posedge clk);
cover forall TAG in {0:15} : // for each tag
  forall C in {0:9} :  // for each time window
     {req && req_tag == TAG; [*C]; ack && ack_tag == TAG};
```

7.14 State tables

Example 7-69 *Verilog* fragment for a 4-way tag comparator

```
module tagcompare4(addr, clk, reset_n, match
  ...);
  parameter ADDRW = 16;        // Width of address.
  parameter TAG   = 6;         // Starting bit of the tag.

  // Tag match interface.
  input  [ADDRW-1:0] addr;
  input              clk, rst_n;
  output [3:0]       match;    // match vector of address against tags.

  // Tag writing interface
  ...

  reg [3:0] vset;                              // The valid bits.
  reg [ADDRW-1:TAG] tag0, tag1, tag2, tag3; // The tags.

  always @(posedge clk or negedge reste_n) begin
    if (!reset_n) begin
      vset <= 4'b0;
      tag0 <= {WIDTH{1'b0}};
      ...
  end
// how does vset get set to a value? This is confusing.
  assign match = { // create vector of valid and matching.
                   vset[3] && (addr[ADDRW-1:TAG] == tag3),
                   vset[2] && (addr[ADDRW-1:TAG] == tag2),
                   vset[1] && (addr[ADDRW-1:TAG] == tag1),
                   vset[0] && (addr[ADDRW-1:TAG] == tag0)
                 };
  ...
```

Context. Hardware state tables are typically containers that are used to hold sets of specific data that is referenced at some future point in time (for example, a valid address associated with a tag in a cache system). These state tables are often a part of designs containing lookup tables, such as a request bus interfaces or hardware cache coherency algorithm. To function correctly, these tables have specific rules related to correct state of the information they are holding. The rules fall into two categories:

- Table entries must contain legal state.
- Composition of table entries must contain legal states.

Hardware state tables generally have a lookup function used to retrieve state information (for example, a tag-based or table-index lookup function). We now describe a tag-based lookup function, which is used in the following related discussion on assertions.

N-way tag comparator

Tag comparators are used in queues and state tables to retrieve state information previously stored by an associative key, which is referred to as a *tag*. In Example 7-69 we demonstrate a simple Verilog fragment for a 4-way tag comparator.

7.14.1 Assertion

Pattern name. Multiple tag correct match

Problem. If our n-way tag comparator incorrect matches (or does not match) an comparison address to a tag stored within a state table, the associated data could be corrupted or lost, resulting in unpredictable design behavior.

Motivation. Unlike our simple tag comparator demonstrated in Example 7-69, many tag comparator algorithms can be quite complex, and are prone to error. Hence, state data potentially could be corrupted or lost.

Solution. In Example 7-70, we demonstrate an assertions for an n-way tag comparator. The $insetz() system function is used so that we don't have to write the exact tag comparison statement, that is coded in the RTL. The valid bit is used to decide whether to compare a tag value, or we return a 1'bx, which doesn't match.

Considerations. One other comment about this assertion. We do not recommend that you duplicate your implementation code as part of your assertion. Ideally, the match algorithm should be described dependant of the implementation, and preferably at a high-level of abstraction.

Example 7-70 ***SystemVerilog* N-way tag comparison**

```
property ntag_comp;
  @(posedge clk)
    (tagset_hit == $insetz(addr[ADDRW-1:TAG],
                          vset[0] ? tag0 : 'bx, // Compare valid tag0
                          vset[1] ? tag1 : 'bx, // Compare valid tag1
                          vset[2] ? tag2 : 'bx, // Compare valid tag2
                          vset[3] ? tag3 : 'bx  // Compare valid tag3
                           ));
endproperty
assert property (ntag_comp);
```

7.15 Memories

Context. Memories, a fundamental component within a system, are used to store the instructions and data for processing components. Interfaces to memory consist of control, data, and address buses. The control signals are used to select a specific memory component to service a system read or write request. Assertions added to the interface of memory components helps isolate modeling problems quickly; such as illegal control signal combinations—or illegal control, address, and data signal values (such as X). Example 7-71 demonstrates a Verilog fragment for a memory module interface, which we use as a reference for the assertions specified in this section.

Example 7-71 Verilog fragment for a memory module interface

```
module memory_device(ce_n, rd_n, we_n, addr, wdata, data, clk);
  input          ce_n,    // chip select ( -- _n == active low )
                 rd_n,    // read enable
                 we_n,    // write enable
                 clk;     // the clock
  input   [11:0] addr;    // the address for the device
  input    [7:0] wdata;   // write data for writes
  output   [7:0] data;    // read data being returned.
```

7.15.1 Assertions

7.15.1.1 Unknown controls

Pattern name. Unknown controls

Problem. If a neighboring block within an RTL model drives X values as controls into a memory component, unpredictable behavior can occur, which could be difficult to isolate if the X value is not visible at the chip-level boundary.

we encourage using lint to identify unconnected ports

Motivation. When an engineer instantiates a memory model, it is possible that the controls signals were inadvertently left unconnected, which would result in X values driven into the memory model during simulation. Furthermore, neighboring components could source an X value into the memory model due to an internal error, which might create unpredictable behavior. Hence, trapping illegal values on major block interfaces reduces debug time by isolating problems closer to their source.

Solution. Example 7-72 specifies an assertion that checks for unknown control signals due to unconnected ports or neighboring components sourcing an illegal X value.

Example 7-72 ***SystemVerilog* check for illegal control signals.**

```
assert property (@(posedge clk) not $isunknown(ce_n, rd_n, we_n))
   else $error("Unknown control signals present (%0b).",
                {ce_n, rd_n, we_n});
```

7.15.1.2 Unknown address

Problem. If a neighboring block within an RTL model drives X values as address into a memory component, unpredictable behavior can occur, which could be difficult to isolate if the X value is not visible at the chip-level boundary.

we encourage using lint to identify unconnected ports

Motivation. When an engineer instantiates a memory model, it is possible that address signals were inadvertently left unconnected, which would result in X values driven into the memory model during simulation. Furthermore, neighboring components could source an X value into the memory model due to an internal error, which might create unpredictable behavior. Hence, trapping illegal values on major block interfaces reduces debug time by isolating problems closer to their source.

Solution. Example 7-73 specifies an assertion that checks for unknown address signals due to unconnected ports or neighboring components sourcing an illegal X value.

Example 7-73 ***SystemVerilog* address with request is legal**

```
property legal_request;
  @(posedge clk)
  not (~ce_n & (~rd_n | ~we_n ) |-> $isunknown(addr));
endproperty
assert property (legal_request)
   else $error("Unknown address during memory request.");
```

7.15.1.3 Unknown store data

Problem. If a neighboring block within an RTL model drives X values as data into a memory component, unpredictable behavior can occur, which could be difficult to isolate if the X value is not visible at the chip-level boundary.

we encourage using lint to identify unconnected ports

Motivation. When an engineer instantiates a memory model, it is possible that data signals were inadvertently left unconnected, which would result in X values driven into the memory model during simulation. Furthermore, neighboring components could source an X value into the memory model due to an internal error, which might create unpredictable behavior. Hence, trapping illegal values on major block interfaces reduces debug time by isolating problems closer to their source.

Solution. Example 7-74 specifies an assertion that checks for unknown data signals due to unconnected ports or neighboring components sourcing an illegal X value.

Example 7-74 ***SystemVerilog*** **store data is known**

```
property storedata_unknown;
  @(posedge clk) not (~ce_n & ~we_n |-> $isunknown(wdata));
endproperty
assert property (storedata_unknown)
   else $error("Unknown write data presented for operation.");
```

7.15.1.4 Unknown read data

Pattern name. Unknown read data

Problem. If a neighboring block within an RTL model reads an X value as data from a memory component, unpredictable behavior can occur, which could be difficult to isolate if the X value is not visible at the chip-level boundary.

we encourage using lint to identify unconnected ports

Motivation. If a memory component was not initialized properly, or incorrectly addressed, a neighboring components could read an X value, which might create unpredictable behavior. Hence, trapping illegal values on major block interfaces reduces debug time by isolating problems closer to their source.

Solution. Example 7-74 specifies an assertion that checks for unknown data read from a memory.

Example 7-75 ***SystemVerilog*** **read data is known**

```
property readdata_unknown;
  @(posedge clk) not (~ce_n & ~rd_n |-> $isunknown(data));
endproperty
assert property (readdata_unknown)
   else $error("Unknown data returned for read operation.");
```

7.16 Arbiter

Context. Arbiters are a critical component in systems containing shared resources. For example, a system containing multiple processors that share a as a common memory bus would require an arbitration scheme to prevent multiple processors accessing the bus at the same time. The are a number of different arbitration schemes, such as the unfair priority scheme or the fair round-robin scheme, which are not discussed in this book. In this section, we demonstrate a few common assertions, which are useful at identifying problems in the arbiter's implementation. Example 7-76 demonstrates a simple arbiter interface, which is used in the following related discussion on assertions.

This arbiter chooses between four requestors. A requestor corresponds to a single bit in each signal (the same bit position.) A requestor may make a request by asserting its bit in the vector (for example `req[0]`). When it receives the corresponding grant (for example, gnt[0]) it is allowed access to the bus. The `hipri` signal is used by the requestor to signal a high priority request. It asserts both its request bit and its high priority bit (for example req[0], hipri[0]) to the arbiter. To simplify our discussion, it is expected that only one high priority request is asserted, if any.

Example 7-76 ***Verilog*** **fragment for simple arbiter interface**

```
module simple_arb(
 input  [3:0] req;   // Request vector.
 input  [3:0] hipri; // High priority flags.
 output [3:0] gnt;   // Grant vector.

 // Grant is asserted two cycles (minimum)
 // after request is asserted.
  ...
```

7.16.1 Assertions

7.16.1.1 Request timeout

Pattern name. Request timeout

Problem. Failure to generate a grant to a requestor in a reasonable period of time often indicates a a deadlock situation.

Motivation. Forward progress problems, related to bus transaction, can be difficult to identify during simulation. By adding performance restrictions on the servicing of a request to an arbiter, we can isolate deadlock situations closer in time (and location) to the error in the design.

Solution. Example 7-77 specifies a performance limit for the service of a request, which states that a `grant` must be received within 50 cycles of a `req`.

Example 7-77 ***SystemVerilog* req receives a grant within N cycles**

```
assert property (@(posedge clk) $rose(req) |-> ##[2:50] grant)
  else $error("Request did not receive grant within timout limit.");
```

7.16.1.2 High priority request

Pattern name. High priority request

Problem. If a priority arbitration scheme is violated, a critical system event might not be serviced—causing unpredictable behavior that can be difficult to debug or isolate at the chip-level during verification.

Motivation. High priority arbitration schemes must ensure that the highest priority requestor immediately receives the next grant after the current transaction completes, otherwise a critical system event could go unserviced.

Solution. Example 7-78 demonstrates how to specify an assertion for a priority arbitration scheme. Note that we use the Verilog-2001 generate construct. The generate construct is similar to the PSL forall operator, which enables us to iterate through the various priority levels for a grant. This example asserts that for all request associated with a priority level, then if the request was the highest priority, a grant associated with that request must be generated within 50 cycles—and no other lower priority grant can be generated before the highest priority grant (which is checked by the $stable system functional).

Example 7-78 *SystemVerilog* high priority req's receive grant next

```
genvar i;
// Property to detect a high priority request and then expect
// the next grant change to be for this requestor.
property hipri_grant(N)
    @(posedge clk) ($rose(req[N] && hipri[N])
      |=> ($stable(grant) [*1:50] ##1 grant[N]))
endproperty

// Generate an assertion for each requestor.
generate for(i=0;i<4;i = i + 1)
  assert property (hipri_grant(i))
    else $error("Grant[%d] for hi priority not next or timedout.", i);
endgenerate
```

7.16.1.3 Round-robin arbitration

Pattern name. Round-robin arbitration

Problem. If a round-robin arbitration scheme is violated, than a particular requester could go un-serviced.

Motivation. Round-robin arbitration schemes are fair, in that a requester cannot receive a grant if any other request has been received, The arbiter will generate a grant in a circular or rotation fashion to a neighboring request from the list. If the arbitration scheme were in error and skipped a requestor, then potentially *starvation* could occur, which means that the requestor would never be serviced.

Solution. In Example 7-79 we demonstrate how to specify assertions for a round-robin arbiter, to ensure that the scheme is fair (that is, no single requestor could be serviced multiple times if another request has been made—and the next request is the immediate neighboring request from a list). This assertion is written for consecutive pairs of requests (N and N+1.) It can be instantiated for each pair ((0, 1) (1, 2) (2, 3)) of requests. A different assertion could be written for nonconsecutive pairs.

Example 7-79 ***SystemVerilog* round-robin arbitration assertion**

```
// Request N+1 needs to be asserted 2 cycles before end of current
// transaction.

// Property to expect the next requestor is granted if they are
// requesting at the appropriate time before the transaction end.

property arb_rotation(N);
  @(posedge clk) (req[N+1] ##1
    req[N+1] & req[N] & gnt[N] & end_of_trans |=> gnt[N+1]);
endproperty

assert property (arb_rotation(0))
    else $error("Grant did not switch to next request at end of transaction.");

// Next combination of request(1, 2) and current grant...
assert property (arb_rotation(1))
    else $error("Grant did not switch to next request at end of transaction.");
...
```

7.16.2 Functional coverage

7.16.2.1 Grant transition

Pattern name. Grant transition

Problem. If we do not adequately explore all possible combination of grants generated by an arbiter, then potentially a corner case error will go untested.

Motivation. Often, corner case bugs are related to complex combination of events, which are unexpected. The completion of one transaction may affect the next transaction after a specific period. Hence, it is important to identify which combination of grants generated by an arbiter have been explored.

Solution. Example 7-80 reports the occurrence of any pairing of back-to-back grants by an arbiter, which helps us determine which pairs have not been explored.

Example 7-80 ***PSL* grant transfers between all pairs of requestors**

```
cover forall R1 in {0:3} :
  forall R2 in {0:3} :
    {gnt[R1]; gnt[R2]};
```

7.17 Summary

In this chapter, we explored a typical set of assertions and functional coverage points for queues, stacks, finite state machines, encoders, decoders, multiplexers, state table structures, memory, and arbiters. From this base set of assertions, we expect you would create additional assertions to cover your specific needs. Our experience has been during the process of specifying assertions and functional coverage, errors are detected and corrected. Furthermore, during verification the time spent debugging problems within your designs is significantly reduced

APPENDIX

OPEN VERIFICATION LIBRARY

The Accellera Open Verification Library [Accellera OVL 2003] provides designers, integrators, and verification engineers with a single, vendor-independent assertion standard interface for design validation using simulation, semi-formal verification, and formal verification techniques. By using a single, well-defined interface, the Open Verification Library bridges the gap between different types of verification, and makes more advanced verification tools and techniques available to non-expert users.

The Open Verification Library is composed of a set of Verilog and VHDL assertion monitors that are defined by the Accellera Open Verification Library committee. This set of monitors enables the designer to check specific properties of a design.

In this section, we discuss thirteen of the most popular OVL monitors. For a complete listing of Accellera Open Verification Library monitors, see *www.verificationlib.org*.

A.1 OVL methodology advantages

The Accellera Open Verification Library (OVL) assertion monitors provide many systematic elements for an effective assertion-based verification methodology, which are typically not addressed by general property languages. For instance, the OVL incorporates a consistent and systematic means of specifying RT-level implementation assertions structurally through a set of concurrent assertion monitors. These monitors provide designers with a module, which guides them to express a broad class of assertions. In addition, these monitors address methodology considerations by providing uniformity and predictability within an assertion-based verification flow and encapsulating the following features:

- unified and systematic method of reporting that can be customized per project
- common mechanism for enabling and disabling assertions during the verification process
- systematic method of filtering the reporting of a specific assertion violation by limiting the firing report to a configured amount

Finally, the OVL does not require a pre-processor to take advantage of assertion specification in the RTL source. Furthermore, the designer does not have to wait until EDA vendors provide tool support for emerging standards since the library is written in standard IEEE-1364 Verilog and IEEE-1076 VHDL. In other words, it will work *right out of the box* for today's designs. This means that IP containing assertions can be delivered to customers without the need to deliver any addition tools for preprocessing the assertion into simulation monitors.

A.2 OVL standard definition

All assertion monitors defined by the Open Verification Library initiative observe the following BNF format, defined in compliance with Verilog Module instantiation of the IEEE Std 1364-1995 "Verilog Hardware Description Language".[1]

```
assertion_instantiation ::= assert_identifier
  [parameter_value_assignment] module_instance ;

parameter_value_assignment ::= #(severity_level
  {,other parameter expressions}, options, msg)

module_instance ::= name_of_instance
  ([list_of_module_connections])

name_of_instance ::= module_instance_identifier

list_of_module_connections ::=
   ordered_port_connection
```

1. In this appendix, we provide a formal definition for the Verilog version of the OVL. The definition for the VHDL version of the library can be obtained at [Accellera OVL 2003].

```
    {,ordered_port_connection}
  | named_port_connection {,named_port_connection}

ordered_port_connection ::= [expression]

named_port_connection ::= .port_identifier
   ([expression])

assert_identifier ::= assert_[type_identifier]

type_identifier ::= identifier
```

A.2.1 OVL runtime macro controls

The Assertion Monitor Library currently includes four Verilog Macro Global Variables:

- `ASSERT_GLOBAL_RESET
- `ASSERT_MAX_REPORT_ERROR
- `ASSERT_ON
- `ASSERT_INIT_MSG

These four variables are described briefly in the table below and in greater detail in the following paragraphs.

Variable	**Definition**
ASSERT_GLOBAL_RESET	Overrides individual *reset_n* signals
ASSERT_MAX_REPORT_ERROR	Defines the number of errors required to trigger a report
ASSERT_ON	Enables assertion monitors during verification
ASSERT_INIT_MSG	Prints a report that lists the assertions present in a given simulation environment.

The `list_of_module_connections` has one required parameter, `reset_n`. The signal `reset_n` is an active low signal that indicates to the assertion monitor when the initialization of the circuit being monitored is complete. During the time when *reset_n* is low, the assertion monitor will be disabled and initialized. Alternatively, to specify a reset_n signal or condition for each assertion monitor, you may specify the global macro variable `ASSERT_GLOBAL_RESET. If this variable is defined, all instantiated monitors will disregard their respective *reset_n* signals. Instead, they will be initialized whenever `ASSERT_GLOBAL_RESET is low.

Every assertion monitor maintains an internal register *error_count* that stores the number of times the assertion monitor instance has fired. This internal register can be accessed by the testbench to signal when a given testbench should be aborted. When the global macro variable `ASSERT_MAX_REPORT_ERROR is defined, the assertion instance will stop reporting messages if the number of errors for that instance is greater than the value defined by the `ASSERT_MAX_REPORT_ERROR macro.

To enable the assertion monitors during verification, you must define the macro 'ASSERT_ON (for example, +define+ASSERT_ON). During synthesis, the ASSERT_ON would not be defined. In addition, *//synthesis translate_off* meta-comments are contained within the body of each monitor to prevent accidental synthesis of the monitor logic.

When you define the `ASSERT_INIT_MSG macro, an "initial" block calls a task to report the instantiation of the assertion. This macro is useful for identifying each of the assertions present in a given simulation environment.

Please note: Most assertions are triggered at the positive edge of a triggering signal or expression `clk`. The assertion assert_proposition is an exception, it monitors an expression at all times.

A.2.2 Customizing OVL messages

The OVL, which is available online at *www.verificationlib.org,* includes a file named *ovl_task.h,* as shown in Example A-1. This file contains a set of tasks that allow you to customize the following:

- simulation startup identification of assertions
- error message reporting mechanism
- actions associated with assertion firing (for example, $finish)

You may use this file to call your own PLI user-defined task upon triggering an assertion, as an alternative to the default $display reporting mechanism currently built into the OVL. To take advantage of this feature, simply edit the *ovl task.h* file shown below to reflect your preferences.

Example A-1 Definition for Verilog version of *ovl_task.h*

```
task ovl_error;
    input [8*63:0] err_msg;
  begin

`ifdef ASSERT_MAX_REPORT_ERROR
    if (error_count <= `ASSERT_MAX_REPORT_ERROR)
`endif
    if (severity_level == 0) begin
      error_count = error_count + 1;
      $display("OVL_FATAL : %s : %s : %0s : \ severity %0d : time %0t : %m",
        assert_name, msg, err_msg,
        severity_level, $time);
      ovl_finish;
    end

  else if (severity_level == 1) begin
      $display("OVL_ERROR : %s : %s : %0s : \ severity %0d : time %0t : %m",
        assert_name, msg, err_msg,
        severity_level, $time);
      error_count = error_count + 1;
    end

    else if (severity_level == 2) begin
      $display("OVL_WARNING : %s : %s : %0s : \ severity %0d : time %0t : %m",
        assert_name, msg, err_msg,
        severity_level, $time);
      ovl_warning;
    end

    else begin
      if ((severity_level > 4) ||
          (error_count == 1))
        $display("OVL_NOTE : %s : %s : %0s : \ severity %0d : time %0t : %m",
          assert_name, msg, err_msg,
          severity_level, $time);
    end

  end
endtask
task ovl_finish;
  begin
    #100 $finish;
  end
endtask
```

Example A-1 Definition for Verilog version of *ovl_task.h*

```
task ovl_warning;
  begin
      // Some user defined Stuff Here, e.g., PLI
  end
endtask

task ovl_init_msg;
  begin
    $display("OVL_NOTE: %s initialized @ %m \ Severity: %0d, Message: %s",
          assert_name,severity_level, msg);
  end
endtask
```

A.3 Firing OVL monitors

During simulation, an OVL assertion monitor will "fire" (that is, report an error) when the specific check performed by the OVL monitor detects an error, generally on the rising edge of the user-supplied sample clock. The ovl_error() task is called to report the assertion violation.

However, when the expression being asserted (that is, checked) is not synchronized with the assertion sample clock (for example, due to a race condition) either a non-deterministic triggering delay or false assertion firing may occur.

- *Non-deterministic triggering delay* refers to the delay between the time the error condition occurs and the time it is detected.
- *False firing* can occur with more complex assertions if the *test_expr* and assertion sample clock are not synchronized properly.

To avoid these consequences, consider an appropriate assertion monitor sampling clock *clk* related to the test_expr. Furthermore, using a variable derived from non-blocking assignments within the `test_expr` greatly minimizes the possibility of race conditions. Experience has shown that most false firings of assertions that are a result of race conditions, are due to signals originating from the testbench or from blocking assignments within the RTL code.

A.4 Using OVL assertion monitors

The OVL set of assertion monitors can be used to improve design verification concerns. In general, follow the guidelines listed below when making decisions about placement for assertion monitors in your RTL code:

- Include assertion monitors to capture all design assumptions during RTL coding, or corner case concerns.
- Include assertion monitors when a module has an external interface. In this case, assumptions on the correct input and output behavior should be guarded and verified.
- Include assertion monitors when interfacing with third party modules, since the designer may not be familiar with the module description (as in the case of IP cores) or may not completely understand the module. In these cases, guarding the module with assertion monitors may prevent incorrect use of the module.

Usually, a specific assertion monitor is suited to cover a desired property, which poses a potential problem. In other cases, even though a specific assertion monitor may not exist, a combination of two or three assertion monitors that perhaps include some additional RTL code (for example, bracketed by 'ifdef ASSERT_ON in Verilog) provide the desired coverage.

The number of actual assertions you must add to a specific design varies. A design might require a few or hundreds, depending on the complexity of the design and the complexity of the properties that must be checked.

Writing assertion monitors for a given design requires careful analysis and planning for maximum efficiency. While writing too few assertions may not increase the coverage on a design, writing too many assertions may increase verification time, sometimes without increasing the coverage. In most cases, however, the runtime penalty incurred by adding assertion monitors is relatively small. The significant reduction in debug time provided by an assertion-based methodology more than compensates for the incremental overhead in simulation performance. Hence, the designer should not hesitate to add assertions for all potential corner case concerns within the design. Without capturing these potential problem points or design assumptions within the implementation, there is little chance that these corner case concerns will be validated by the verification team (which is generally focused on verifying the design at a higher level of abstraction).

The following sections provide details related to specific OVL monitors for various classes of properties to be checked.

A.5 Checking invariant properties

The OVL provides a set of monitors that check invariant properties. Invariant properties are conditions that must hold, or not hold, for *all* cycles. This section discusses a few of the more common invariant OVL monitors. A complete list of OVL monitors is located at *www.verificationlib.org* [OVL 2002].

A.5.1 assert_always

checking a property that always holds

The OVL assert_always assertion is used to check that an invariant property *always* holds at every clock boundary. For instance, Example A-2 provides a simple demonstration of an OVL assert_always to monitor a unique counter's specified range (that is, 0 to 8).

Example A-2 Monitor legal range for count variable with assert_always

```
module counter_0_to_8 (reset_n,clk,inc,dec);
 input reset_n, clk, inc, dec;
 reg [3:0] count;

 always @(posedge clk) begin
   if (reset_n) count = 4'b0;
   else count = count + inc - dec;
 end
 // OVL check for valid range
 assert_always #(0, 0, 0, "range 0-8 error")
    valid_count (clk, reset_n,
       (count >= 4'b0000) && (count <= 4'b1000));
endmodule
```

Whenever the `inc` signal is TRUE, the counter increments by one. Whenever the `dec` signal is TRUE, the counter decrements by one. For this unique counter, the user must ensure that the controlling logic, which generates the signal `inc` and `dec`, always maintains a specified range for the variable count (that is, 0 to 8).

The syntax for the OVL assert_always monitor is defined as follows:

Syntax ***assert_always [#(severity_level, options, msg)] inst_name (clk, reset_n, test_expr);***

severity_level	Severity of the failure with default value of 0.
options	Vendor options. Currently, the only supported option is *options*=1, which defines the assertion as a constraint on formal tools. The default value is *options*=0, or no options specified.
msg	Error message that will be printed if the assertion fires.
inst_name	Instance name of assertion monitor.
clk	Triggering or clocking event that monitors the assertion.
reset_n	Signal indicating completed initialization (for example, a local copy of reset_n of a global reference to reset_n).
test_expr	Expression being verified at the positive edge of *clk*.

Example A-3 defines the semantics for the OVL assert_always. Note that the *ovl_task.h* definition was previously defined in Example A-1.

Example A-3 Verilog definition for the OVL assert_always

```
module assert_always (clk, reset_n, test_expr);
// synopsys template
  parameter severity_level = 0;
  parameter options = 0;
  parameter msg="VIOLATION";
  input clk, reset_n, test_expr;

//synopsys translate_off
`ifdef ASSERT_ON
  parameter assert_name = "ASSERT_ALWAYS";
  integer error_count;
  initial error_count = 0;

`include "ovl_task.h"

 `ifdef ASSERT_INIT_MSG
    initial
      ovl_init_msg;
  `endif
always @(posedge clk) begin
    `ifdef ASSERT_GLOBAL_RESET
      if (`ASSERT_GLOBAL_RESET != 1'b0) begin
    `else
      if (reset_n != 0) begin  // active low reset
    `endif
        if (test_expr  != 1'b1) begin
          ovl_error("");

        end
      end
  end
`endif
//synopsys translate_on
endmodule
```

A.5.2 assert_never

checking a property that never holds

The OVL assert_never monitor allows us to specify an invariant property that should *never* evaluate to TRUE. For instance, if we modify Example A-2 and specify that the count variable should never be greater than 8, then we can use the OVL assert_never monitor as shown in Example A-4 to check this condition.

Example A-4 Monitor legal range for count variable with assert_never

```
module counter_0_to_8 (reset_n,clk,inc,dec);
 input reset_n, clk, inc, dec;
 reg [3:0] count;

 always @(posedge clk) begin
   if (reset_n) count = 4'b0;
   else count = count + inc - dec;
 end
 // OVL check for valid range
 assert_never #(0, 0, 0, "range 0-8 error")
   valid_count (clk, reset_n, (count > 4'b1000));
endmodule
```

The syntax for the OVL assert_never monitor is defined as follows:

Syntax

assert_never [#(severity_level, options, msg)] inst_name (clk, reset_n, test_expr);

severity_level	Severity of the failure with default value of 0.
options	Vendor options. Currently, the only supported option is *options*=1, which defines the assertion as a constraint on formal tools. The default value is *options*=0, or no options specified.
msg	Error message that will be printed if the assertion fires.
inst_name	Instance name of assertion monitor.
clk	Triggering or clocking event that monitors the assertion.
reset_n	Signal indicating completed initialization (for example, a local copy of *reset_n* of a global reference to *reset_n*).
test_expr	Expression being verified at the positive edge of *clk*.

Note that the parameter and argument options specified for the assert_never module are the same as the options used in the assert_always monitor. Also, the semantic definition for assert_never is similar to the definition for assert_always. However, the assert_never monitor checks that `test_expr` is *never* equal to 1'b1, while the assert_always monitor checks that `test_expr` is *always* equal to 1'b1.

A.5.3 assert_zero_one_hot

checking for mutually exclusive events

The assert_always and the assert_never monitors are convenient for specifying general invariant properties. However, there are many specific invariant properties where formulating the Boolean expression is cumbersome or awkward. Examples include: checking for one-hot or one-cold conditions, even or odd parity, or valid variable range conditions.

For instance, Example A-5 illustrates how we could check that the individual bits contained within the `cntrl` variable are mutually exclusive. For this assertion, we use the mathematical property:

```
(cntrl & (cntrl-1))
```

which will always equal zero if the `cntrl` bit vector variable is either all zeroes—or only one bit of the variable is active high at a time. We could check this condition using an assert_always monitor and our mathematical equation, as shown in Example A-5.

Example A-5 Check for zero or one hot condition on cntrl variable

```
module pass_mux (clk,reset_n,cntrl,in,out);
 input        clk, reset_n;
 input [3:0] cntrl,
 input [3:0] in;
 output       out;
 reg          out;
 always @(posedge clk) begin
   case (cntrl)
     4'b0001: out <= in[0];
     4'b0010: out <= in[1];
     4'b0100: out <= in[2];
     4'b1000: out <= in[3];
   endcase
 end
// check for zero or one-hot values for cntrl
assert_always valid_cntrl (clk, reset_n,                  ((cntrl &
(cntrl-1)) == 4'b0));
endmodule
```

However, as previously stated, checking for a zero or one hot condition as demonstrated in Example A-5 is awkward.

Alternatively, the OVL assert_zero_or_one_hot monitor, as shown in Example A-6, simplifies our coding effort while explicitly stating the property we are asserting.

Example A-6 Using assert_zero_one_hot to check cntrl

```
assert_zero_one_hot #(0, 4) valid_cntrl (clk,
                                  reset_n, cntrl);
```

The syntax for the OVL assert_zero_one_hot monitor is defined as follows:

Syntax ***assert_zero_one_hot [#(severity_level, width, options, msg)] inst_name (clk, reset_n, test_expr);***

severity_level	Severity of the failure with default value of 0.
width	Width of the monitored expression *test_expr*.
options	Vendor options. Currently, the only supported option is *options*=1, which defines the assertion as a constraint on formal tools. The default value is *options*=0, or no options specified.
msg	Error message that will be printed if the assertion fires.
inst_name	Instance name of assertion monitor.
clk	Triggering or clocking event that monitors the assertion.
reset_n	Signal indicating completed initialization (for example, a local copy of reset_n of a global reference to reset_n).
test_expr	Expression being verified at the positive edge of *clk*.

The assert_zero_one_hot assertion is most useful in control circuits. It ensures that the state variable of a finite-state machine (FSM) implemented with zero-one-hot encoding will maintain proper behavior. In data path circuits, assert_zero_one_hot can be used to ensure that the enabling signals of bus-based designs will not generate bus contention. Examples of uses for assert_zero_one_hot include controllers, circuit enabling logic, and arbitration logic. Finally, assert_zero_one_hot is useful for checking mutual exclusivity between multiple events. For instance, if fsm_1, fsm_2, and fsm_3 are not permitted to be in a WR_MODE at the same time, then we could write an assertion as follows using the Verilog concatenation operator to create a bit vector that must either be zero or one-hot:

Example A-7 Ensure WR_MODE mutual exclusivity between multiple FSMs

```
assert_zero_one_hot #(0,3) wr_mode (clk, reset_n,
         {fsm_1=='WR_MODE,
          fsm_2=='WR_MODE,
          fsm_3=='WR_MODE});
```

A.5.4 assert_range

checking valid ranges

The OVL assert_range continuously monitors the *test_expr* at every positive edge of the triggering event or clock *clk*. It asserts that a specified Verilog expression will always have a value within a legal *min/max* range, otherwise, a monitor will fire (that is, an error condition will be detected in the Verilog code). The *test_expr* can be either a valid Verilog wire or reg variable, or any valid Verilog expression. The *min* and *max* should be a valid parameter and *min* must be less than or equal to *max*.

Syntax

assert_range [#(severity_level, width, min, max, options, msg)] inst_name (clk, reset_n, test_expr);

severity_level	Severity of the failure with default value of 0.
width	Width of the monitored expression *test_expr*.
min	Minimum value allowed for range check. Default to 0.
max	Maximum value allowed for range check. Default to (2***width* - 1).
options	Vendor options. Currently, the only supported option is *options*=1, which defines the assertion as a constraint on formal tools. The default value is *options*=0, or no options specified.
msg	Error message that will be printed if the assertion fires.
inst_name	Instance name of assertion monitor.
clk	Triggering or clocking event that monitors the assertion.
reset_n	Signal indicating completed initialization (for example, a local copy of reset_n of a global reference to reset_n).
test_expr	Expression being verified at the positive edge of *clk*.

The assert_range assertion should be used in circuits to ensure the proper range of values in control structures, such as counters and finite-state machines (FSM). In datapath circuits, this assertion can be used to check whether the variable or expression is evaluated within the allowed range.

For instance, Example A-8 demonstrates a counter whose range must remain between 1 and 9. The assertion ensures that the controlling logic generating the inc and dec commands maintains a valid range for this counter.

Example A-8 Monitor legal range for count variable with assert_never

```
module counter_1_to_9 (reset_n,clk,inc,dec);
 input reset_n, clk, inc, dec;
 reg [3:0] count;

 always @(posedge clk) begin
   if (reset_n) count = 4'b1;
   else count = count + inc - dec;
 end
 // OVL check for valid range 1 thru 9
 assert_range #(0,4,1,9) count_range_check (clk,     reset_n, count);
endmodule
```

A.6 Checking cycle relationships

In the previous section, we discussed the OVL invariant class of assertions. While specifying properties that must hold (or never hold) for all cycles is certainly important, there are times when it is necessary to be more specific about the correct cycle relationship between multiple events over time. In this section, we introduce a set of OVL monitors that permit us to assert (that is, validate) correct cycle relationships.

A.6.1 assert_next

The OVL assert_next assertion validates proper cycle timing relationships between two events in the design. For instance, whenever event A occurs, then event B must occur on the following cycle (that is, A -> next B). In this implication, event A is referred to as the *antecedent*, while event B is referred to as the *consequence*. For the OVL monitor, the antecedent is represented by the *start_event* expression, while the consequence is represented by the *test_expr* expression.

In addition to advancing time by one cycle after the occurrence of *start_event* for the check of *test_expr* (that is, event B must occur exactly one cycle after event A) it is possible to specify that the *test_expr* will hold multiple clock cycles after the occurrence of the *start_event* expression. This can be specified through the OVL *num_cks* parameter.

Syntax ***assert_next [#(severity_level, num_cks, check_overlapping, only_if, options, msg)] inst_name (clk, reset_n, start_event, test_expr);***

severity_level	Severity of the failure with default value of 0.
num_cks	The number of clocks *test_expr* must evaluate to TRUE after *start_event* is asserted.
check_overlapping	If 1, permits overlapping sequences. In other words, a new *start_event* can occur (starting a new sequence in parallel) while the previous sequence continues. Default is to permit overlapping sequences (that is, default is 1).
only_if	If 1, a *test_expr* can only evaluate TRUE, if preceded *num_cks* earlier by a *start_event*. If *test_expr* occurs without a *start_event*, then an error is flagged. Default is 0.

options	Vendor options. Currently, the only supported option is *options*=1, which defines the assertion as a constraint on formal tools. The default value is *options*=0, or no options specified.
msg	Error message that will be printed if the assertion fires.
inst_name	Instance name of assertion monitor.
clk	Triggering or clocking event that monitors the assertion.
reset_n	Signal indicating completed initialization (for example, a local copy of reset_n of a global reference to reset_n).
start_event	Starting event that triggers monitoring of the *test_expr*.
test_expr	Expression being verified at the positive edge of *clk*.

An important feature of this assertion is that it supports overlapping sequences (as an option). For instance, if you assert that *test_expr* will evaluate TRUE exactly four cycles after start_event, it is not necessary to wait until the sequence finishes before another sequence can begin.

The assert_next assertion should be used in circuits to ensure a proper sequence of events. Common uses for assert_next are as follows:

- verification that multicycle operations with enabling conditions will always work with the same data
- verification that single-cycle operations operate correctly with data loaded at different cycles
- verification of synchronizing conditions that require that data is stable after a specified initial triggering event (such as in an asynchronous transaction requiring req/ack signals)

For instance, Example A-9 includes two overlapping sequences that are being verified as shown in [1997]. The assertion claims that when 'A' occurs, 'B' will occur exactly 4 cycles later. Notice how a new 'A' (or *starting event*) does occur in this example prior to the completion of the first sequence (that is, *test_expr* 'B'). The assertion would be coded as follows:

Example A-9 assert_next

```
assert_next #(0,4,1) AB_check (clk, reset_n, A, B);
```

Figure A-1 **Overlapping sequences of A's and B's.**

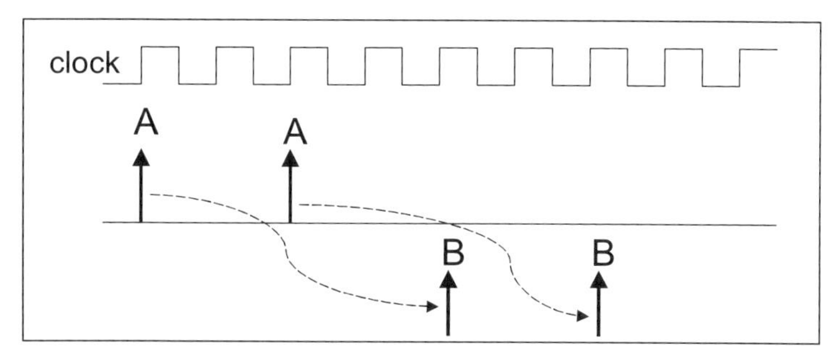

A.6.2 assert_frame

The OVL assert_next monitor allows us to validate an exact cycle relationship between two events within a design. However, often the cycle relationship between multiple events is specified in terms of a range of possibilities (that is, a min/max frame window). The OVL assert_frame assertion validates proper cycle timing relationships within a frame window between two events (that is, Boolean expressions) in the design. When a *start_event* evaluates TRUE, then the *test_expr* must evaluate TRUE within a minimum and maximum number of clock cycles. If the *test_expr* does not occur within these frame window boundaries, an assertion will fire (that is, an error condition will be detected).

The intent of the minimum and maximum range is to identify legal boundaries in which *test_expr* can occur at or after *start_event*. When you specify both the minimum (equal to or greater than 0) and maximum (greater than 0) ranges, a *test_expr* must occur within the specified frame. However, if you do not provide a maximum range, the checker will ensure that the *test_expr* does not occur until after *min_cks*. The frame is from the time of *start_event* through *max_cks*. Additionally, *max_cks* must be greater than *min_cks*.

Special consideration one. If you do not specify a maximum range, the checker ensures that the *test_expr* does not occur until *min_cks* or later. That is, the *test_expr* must not occur before *min_cks*. However, *test_expr* is not required to occur after *min_cks*, we are simply asserting that *test_expr* does not occur before *min_cks*.

Special consideration two. If you specify that *min_cks* is equal to 0, the checker ensures that the *test_expr* occurs prior to *max_cks*. That is, *test_expr* must occur at some point in any cycle from *start_event* through *max_cks*).

Special consideration three. If you specify that both *min_cks* and *max_cks* equal 0, *test_expr* must be true when there is a 0 to 1 transition for *start_event*. That is, *start_event* implies *test_expr* (*start_event* → *test_expr*).

Syntax ***assert_frame [#(severity_level, min_cks, max_cks, flag, options, msg)] inst_name (clk, reset_n, start_event, test_expr);***

severity_level	Severity of the failure.
min_cks	The *test_expr* cannot occur prior to (but not including) the specified minimum number of clock cycles. That is, if *test_expr* occurs at or after *start_event* but before *min_cks*, then an error occurs. **The exception is:** When *min_cks* is set to 0, then there is no minimum check (that is, *test_expr* may occur at start event).
max_cks	The *test_expr* must occur at or prior to the specified number of clock cycles. That is, if the *test_expr* does not occur at or prior to *max_cks*, then an error occurs. **The exception is:** When *max_cks* is set to 0, then there is no maximum check (any value is valid).
flag	0 - Ignores any asserted *start_event* after the first one has been detected (default); 1 - Re-start monitoring *test_expr* if *start_event* is asserted in any subsequent clock while monitoring *test_expr*; 2 - Issue an error if an asserted *start_event* occurs in any clock cycle while monitoring *test_expr*.
options	Vendor options. Currently, the only supported option is *options*=1, which defines the assertion as a constraint on formal tools. The default value is *options*=0, or no options specified.
msg	Error message that will be printed if the assertion fires.
inst_name	Instance name of assertion monitor.
clk	Triggering or clocking sampling event for assertion.
reset_n	Signal indicating completed initialization (for example, a local copy of reset_n of a global reference to reset_n).
start_event	Starting event that triggers monitoring of the *test_expr*.
test_expr	Expression being verified at the positive edge of *clk*.

The assert_frame assertion should be used in control circuits to ensure proper synchronization of events. Common uses of assert_frame are as follows:

- verification that multicycle operations with enabling conditions will always work with the same data

- verification of single cycle operations to operate correctly with data loaded at different cycles
- verification of synchronizing conditions that require that data is stable after a specified initial triggering event (such as in an asynchronous transaction requiring req/ack signals)

Example A-10 demonstrates how assert_frame can be used to verify cycle timing relationships between two events. This assertion claims that after the rising edge of `req` is detected, then an `ack` signal must go high within two to four clocks.

Example A-10 assert_frame

```
assert_frame #(0,2,4) check_req_ack (clk,reset_n, req, ack);
```

A.6.3 assert_cycle_sequence

The OVL assert_next and assert_frame enable us to validate the cycle relationship between two events (that is, Boolean expressions that evaluate to TRUE). However, at times we would like to validate a sequence of events. The OVL assert_cycle_sequence can be used to validate the proper occurrence of multiple events within a design.

A contiguous sequence of events is represented by a Verilog (or VHDL) concatenation of multiple Boolean expressions. For instance, in Verilog, the sequence A, followed by B, followed by C, would be expressed as {A,B,C}. This expression is then passed in as the *event_sequence* argument to the assert_cycle_sequence monitor. The size of *event_sequence* is the width of the concatenation expression, which represent the number of clocks required to move forward in time to validate the sequence. Hence, for the *event_sequence* {A,B,C}, the number of clocks (that is, *num_cks*) is 3.

The **assert_cycle_sequence** assertion checks the following:

- If the OVL *necessary_condition* parameter is 0, then this assertion checks to ensure that if all *num_cks-1* prefix (that is, *event_sequence[num_cks-1:1]*) are satisfied, then the last event ("event_sequence[0]") must hold.
- If "necessary_condition" is 1, then this assertion checks that once the first event ("event_sequence[num_cks-1]") holds, all the remaining events must be satisfied.

Syntax ***assert_cycle_sequence [#(severity_level, num_cks, necessary_condition, options, msg)] inst_name (clk, reset_n, event_sequence);***

severity_level	Severity of the failure with default value of 0.
num_cks	The width of the *event_sequence* (length of number of clock cycles in the sequence) that must be valid. Otherwise, the assertion will fire; that is, an error occurs.
necessary_condition	Either 1 or 0. The default is 0.
options	Vendor options. Currently, the only supported option is *options*=1, which defines the assertion as a constraint on formal tools. The default value is *options*=0, or no options specified.
msg	Error message that will be printed if the assertion fires.
inst_name	Instance name of assertion monitor.
clk	Triggering or clocking event that monitors the assertion.
reset_n	Signal indicating completed initialization (for example, a local copy of reset_n of a global reference to reset_n).
event_sequence	A Verilog concatenation expression, where each bit represents an event.

The assert_cycle_sequence assertion should be used in circuits to ensure a proper sequence of events. An event is a Verilog expression that evaluates TRUE. Common uses for assert_cycle_sequence are as follows:

- verification that multicycle operations with enabling conditions will always work with the same data
- verification of single cycle operations to operate correctly with data loaded at different cycles
- verification of synchronizing conditions that require that data is stable after a specified initial triggering event (such as in an asynchronous transaction requiring req/ack signals)

The following example asserts that when write cycle starts, followed by one wait statement, then the next opcode will have the value `'DONE`.

Example A-11 assert_cycle_sequence

```
assert_cycle_sequence #(0,3)  init_test (clk, reset_n,
      {r_opcode == 'WRITE, r_opcode == 'WAIT,
       r_opcode == 'DONE});
```

A.7 Checking event bounded windows

Many temporal properties are bounded by events (that is, Boolean expressions). For instance, often the designer would like to check that after a *start_event* occurs, then a specific Boolean expression must change values prior to the occurrence of an *end_event*. Figure A-2 illustrates this idea. For this example, the *start_event* is p, while the *end_event* is r. Hence, the Boolean expression q must change values (represented by the color change) after p and before r.

Figure A-2 **Event bounded window for a liveness property**

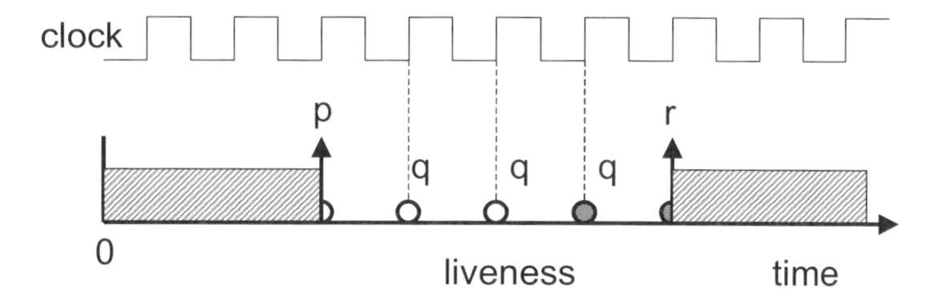

Alternatively, the designer might check that the specified Boolean expression must not change values in an event bounded window.

A.7.1 assert_win_change

The assert_win_change assertion continuously monitors the *start_event* at every positive edge of the triggering event or clock *clk*. When this signal (or expression) evaluates TRUE, the assertion monitor ensures that the *test_expr* changes values prior to the *end_event*.

When the *start_event* evaluates TRUE, the assertion monitor ensures that the *test_expr* changes values on a clock edge at some point up to and including the next *end_event*. Once the *test_expr* evaluates TRUE, it is not necessary for it to remain TRUE throughout the remainder of the test (up to and including *end_event*). Hence, the *test_expr*, does not have to be TRUE at the *end_event*, provided it was true at some point during the test (up to and including *end_event*).

Syntax ***assert_win_change [#(severity_level, width, options, msg)] inst_name (clk, reset_n, start_event, test_expr, end_event);***

severity_level	Severity of the failure with default value of 0.
width	Width of *test_expr* with default value of 1.
options	Vendor options. Currently, the only supported option is *options*=1, which defines the assertion as a constraint on formal tools. The default value is *options*=0, or no options specified.
msg	Error message that will be printed if the assertion fires.
inst_name	Instance name of assertion monitor.
clk	Triggering or clocking event that monitors the assertion.
reset_n	Signal indicating completed initialization (for example, a local copy of reset_n of a global reference to reset_n).
start_event	Starting event that triggers monitoring of the *check_expr*
test_expr	Expression being verified at the positive edge of *clk*.
end_event	Event that terminates monitoring of a *start_event.*

The assert_win_change assertion should be used in circuits to ensure that after a specified initial event, a particular variable or expression will change after the *start_event* and before the *end_event*. Common uses for assert_win_change include:

- verification that synchronization circuits respond after a specified initial stimuli. For example, that a bus transaction will occur without any bus interrupts. Another example is that a memory write command will not occur while we are in a memory read cycle.
- verification that finite-state machines (FSM) change state or will go to a specific state after a specified initial stimuli and before another specified event occurs

In Example A-12, an assertion checks whether *data_bus* is asserted before an asynchronous read operation is finished.

Example A-12 assert_win_change

```
module processor (clk, reset_n, ..., rd, rd_ack, ...);
  input clk;
  input reset_n;

  output rd
  input rd_ack;

  inout [31:0] data_bus;

  ...

  assert_win_change #(0,32) sync_bus_with_rd
    (clk, reset_n, rd, data_bus, rd_ack);

endmodule
```

A.7.2 assert_win_unchange

The assert_win_unchange assertion continuously monitors the *start_event* at every positive edge of the triggering event or clock *clk*. When this signal (or expression) evaluates TRUE, the assertion monitor ensures that the *test_expr* will not change in value up to and including the *end_event*.

Syntax ***assert_win_unchange [#(severity_level, width, options, msg)] inst_name (clk, reset_n, start_event, test_expr, end_event);***

severity_level	Severity of the failure with default value of 0.
width	Width of *test_expr* with default value of 1.
options	Vendor options. Currently, the only supported option is *options*=1, which defines the assertion as a constraint on formal tools. The default value is *options*=0, or no options specified.
msg	Error message that will be printed if the assertion fires.
inst_name	Instance name of assertion monitor.
clk	Triggering or clocking event that monitors the assertion.
reset_n	Signal indicating completed initialization (for example, a local copy of reset_n of a global reference to reset_n).
start_event	Starting event that triggers monitoring of the *check_expr*
test_expr	Expression being verified at the positive edge of *clk*.
end_event	Event that terminates monitoring of the *start_event*.

The assert_win_unchange assertion should be used in circuits to ensure that after a specified initial event, a particular variable or expression will remain unchanged after the *start_event* and before the *end_event*. Common uses for assert_win_unchange include:

- verification that non-deterministic multicycle operations with enabling conditions will always work with the same data
- verification of synchronizing conditions requiring that data is stable after a specified initial triggering event and before an ending condition takes place.

For instance, a bus transaction will occur without any bus interrupts; or a memory write command will not occur if we are in a memory read cycle.

In Example A-13, assume that for the multi-cycle divider operation to work properly, the value of signal *a* must remain unchanged for the duration of the operation. Completion is signaled by the signal *done*.

Example A-13 assert_win_unchange

```
module division_with_check (clk,reset_n,a,b,start,done);
  input clk, reset_n;
  input [15:0] a,b;
  input start;
  output done;

  wire [15:0] q,r;

  div16 div01 (clk, reset_n, start, a, b, q, r, done);

  assert_win_unchange #(0,16) div_win_unchange_a (clk, reset_n, start,
a, done);

endmodule
```

A.8 Checking time bounded windows

In addition to specifying temporal properties within event bounded windows, many properties are bounded with cycle related time bounded windows. In this section we introduce a set of OVL monitors that allow the design to specify time bounded window assertions.

A.8.1 assert_change

The assert_change assertion continuously monitors the *start_event* at every positive edge of the triggering event or clock *clk*. When the *start_event* evaluates TRUE, the assertion monitor ensures that the *test_expr* changes values (as sampled on a clock edge) at some point within the next *num_cks* number of clocks. As soon as the *test_expr* evaluates TRUE, this assertion is satisfied; and checking is discontinued for the remaining *num_cks* number of clocks.

Syntax

assert_change [#(severity_level, width, num_cks, flag, options, msg)] inst_name (clk, reset_n, start_event, test_expr);

severity_level	Severity of the failure with default value of 0.
width	Width of *test_expr* with default value of 1.
num_cks	The number of clocks for *test_expr* to change its value before an error is triggered after *start_event* is asserted.
flag	0 - Ignores any asserted *start_event* after the first one has been detected (default); 1 - Re-start monitoring *test_expr* if *start_event* is asserted in any subsequent clock while monitoring *test_expr*; 2 - Issues an error if an asserted *start_event* occurs in any clock cycle while monitoring *test_expr*.
options	Vendor options. Currently, the only supported option is *options*=1, which defines the assertion as a constraint on formal tools. The default value is *options*=0, or no options specified.
msg	Error message that will be printed if the assertion fires.
inst_name	Instance name of assertion monitor.
clk	Triggering or clocking event that monitors the assertion.
reset_n	Signal indicating completed initialization (for example, a local copy of reset_n of a global reference to reset_n).
start_event	Starting event that triggers monitoring of the *test_expr*.
test_expr	Expression being verified at the positive edge of *clk*.

The assert_change assertion should be used in circuits to ensure that after a specified initial event, a particular variable or expression will change. Common uses for assert_change include:

- verification that synchronization circuits respond after a specified initial stimuli. For example, in protocol verification, this assertion may be used to check that after a *request* an *acknowledge* will occur within a specified number of cycles.

- verification that finite-state machines (FSM) change state, or will go to a specific state, after a specified initial stimuli.

In Example A-14, the module *synchronizer_with_bug* is designed to respond by asserting *out* after *count_max* cycles *sync* was asserted. Note in the accompanying figure, however, that *out* is not asserted until after the trigger of *clk*. In this figure, the waveform is shown until the error is triggered by the assertion library. At this point, the simulation is aborted.

Example A-14 assert_change

```
module synchronizer_with_bug (clk, reset_n, sync,
  count_max, out);

  input clk, reset_n, sync;
  input [3:0] count_max;
  output out;

  reg out;
  reg [3:0] count;

  always @(posedge clk) begin
    if (reset_n == 0) begin
      out <= 0;
      count <= 0;
    end
    else if (count != 0) begin
      count <= count - 1;
      if (count == 1) out <= 1;
    end
    else if (sync == 1) count <= count_max;
    else if (out == 1) out <= 0;
  end

  // out must change values 3 cycles after sync
  assert_change #(0,1,3,0) synch_test (clk,reset_n,(sync == 1), out);
endmodule
```

A.8.2 assert_unchange

The assert_unchange assertion continuously monitors the *start_event* at every positive edge of the triggering event or clock *clk*. When *start_event* evaluates TRUE, the assertion monitor ensures that the *test_expr* will not change values within the next *num_cks* number of clocks.

Syntax ***assert_unchange [#(severity_level, width, num_cks, flag, options, msg)] inst_name (clk,reset_n, start_event, test_expr);***

severity_level	Severity of the failure with default value of 0.
width	Width of *test_expr* with default value of 1.
num_cks	For this number of clocks after *start_event* is asserted, *test_expr* must remain unchanged. Otherwise, an error is triggered.
flag	0 - Ignores any asserted *start_event* after the first one has been detected (default); 1 - Re-start monitoring *test_expr* if *start_event* is asserted in any subsequent clock while monitoring *test_expr*; 2 - Issue an error if an asserted *start_event* occurs in any clock cycle while monitoring *test_expr*.
options	Vendor options. Currently, the only supported option is *options*=1, which defines the assertion as a constraint on formal tools. The default value is *options*=0, or no options specified.
msg	Error message that will be printed if the assertion fires.
inst_name	Instance name of assertion monitor.
clk	Triggering or clocking event that monitors the assertion.
reset_n	Signal indicating completed initialization (for example, a local copy of reset_n of a global reference to reset_n).
start_event	Starting event that triggers monitoring of the *check_expr*
test_expr	Expression being verified at the positive edge of *clk*.

The assert_unchange assertion should be used in circuits to ensure that after a specified initial event, a particular variable or expression will remain unchanged for some number of clocks. Common uses for assert_unchange include:

- verification that multicycle operations with enabling conditions will always work with the same data
- verification that single cycle operations will operate correctly with data loaded at different cycles
- verification of synchronizing conditions that require data is stable after a specified initial triggering event

In Example A-15, we assume that for the multi-cycle divider operation to work properly, the value of signal *a* must remain unchanged for the duration of the operation that, in this example, is 16 cycles.

Example A-15 assert_unchange

```
module division_with_check (clk,reset_n,a,b,start,done);
  input clk, reset_n;
  input [15:0] a,b;
  input start;
  output done;

  wire [15:0] q,r;

  div16 #(16) div01 (clk, reset_n, start, a, b, q, r, done);

  assert_unchange #(0,16,16) div_unchange_a                    (clk,
reset_n,start == 1, a);
endmodule
```

A.9 Checking state transitions

The OVL provides a set of monitors that can be used to insure proper sequencing of finite state machines (FSM). For example, if an FSM is in a specific state, then the OVL assert_no_transition can be used to monitor an illegal transition.

A.9.1 assert_no_transition

The assert_no_transition assertion continuously monitors the *test_expr* variable at every positive edge of the triggering event or clock *clk*. When this variable evaluates to the value of *start_state*, the monitor ensures that *test_expr* will never transition to the value of *next_state*. The *width* parameter defines the size (that is, number of bits) of the *test_expr*.

Syntax ***assert_no_transition [#(severity_level, width, options, msg)] inst_name (clk, reset_n, test_expr, start_state, end_state);***

severity_level	Severity of the failure with default value of 0.
width	Width of state variable *test_expr* with default value of 1.
options	Vendor options. Currently, the only supported option is *options*=1, which defines the assertion as a constraint on formal tools. The default value is *options*=0, or no options specified.
msg	Error message that will be printed if the assertion fires.
inst_name	Instance name of assertion monitor.

clk	Triggering or clocking event that monitors the assertion.
reset_n	Signal indicating completed initialization (for example, a local copy of reset_n of a global reference to reset_n).
test_expr	State variable representing finite-state machine (FSM) being checked at the positive edge of *clk*.
start_state	Triggering state of *test_expr.*
end_state	Invalid state for the machine represented by *test_expr* when traversed from state *start_state.*

The assert_no_transition assertion should be used in control circuits, especially FSMs, to ensure that invalid transitions are never triggered.

Please note: *start_state* and *end_state* include any valid Verilog expression. As a result, multiple transitions can be specified by encoding the transitions in these variables. Please refer to the following example.

Example A-16 assert_no_transition

```
module counter_09_or_0F                    (reset_n,clk,count,sel_09);
 input reset_n, clk, sel_09;
 output [3:0] count;
 reg [3:0] count;

 always @(posedge clk)
  if (reset_n==0 || count==4'd9 && sel_09==1'b1)
    count <= 4'd0;
  else
    count <= count + 1;
 assert_no_transition #(0,4) valid_count
         (clk, reset_n, count, 4'd9,                   (sel_09 == 1)
? 4'd10 : 4'd0);
endmodule
```

A.9.2 assert_transition

The assert_transition assertion continuously monitors the *test_expr* variable at every positive edge of the triggering event or clock *clk*. When *test_expr* evaluates to the value of *start_state*, the assertion monitor ensures that if *test_expr* changes value, then it will change to the value of *next_state*. The *width* parameter defines the size (that is, number of bits) of the *test_expr.*

Syntax ***assert_transition [#(severity_level, width, options, msg)] inst_name (clk, reset_n, test_expr, start_state, end_state);***

severity_level	Severity of the failure with default value of 0.
width	Width of state variable *test_expr* with default value of 1.
options	Vendor options. Currently, the only supported option is *options*=1, which defines the assertion as a constraint on formal tools. The default value is *options*=0, or no options specified.
msg	Error message that will be printed if the assertion fires.
inst_name	Instance name of assertion monitor.
clk	Triggering or clocking event that monitors the assertion.
reset_n	Signal indicating completed initialization (for example, a local copy of reset_n of a global reference to reset_n).
test_expr	State variable representing finite-state machine (FSM) being checked at the positive edge of *clk*.
start_state	Triggering state of *test_expr.*
end_state	Next valid state for machine represented by *test_expr* when traversed from state *start_state.*

The assert_transition assertion should be used in control circuits, especially FSMs, to ensure that required transitions are triggered.

Please note *start_state* and *end_state* are verification events that can be represented by any valid Verilog expression. As a result, multiple transitions can be specified by encoding the transitions in these variables.

Example A-17 assert_transition

```
module counter_09_or_0F                    (reset_n,clk,count,sel_09);
input reset_n, clk, sel_09;
output [3:0] count;
reg [3:0] count;

always @(posedge clk)
  if (reset_n==0 || count==4'd9 && sel_09==1'b1)
    count <= 4'd0;
  else
    count <= count + 1;

assert_transition #(0,4) valid_count
    (clk, reset_n, count, 4'd9,
         (sel_09 == 1'b0) ? 4'd10 : 4'd0);
endmodule
```

APPENDIX

B

PSL PROPERTY SPECIFICATION LANGUAGE

B.1 Introduction to PSL

PSL (Property Specification Language) is a formal property language created by Accellera that is compatible with any HDL. This language was based on the Sugar property language originally developed by IBM. A specification written in PSL is both easy to read and mathematically precise, which makes it ideal for both documentation and verification. Thus, it is ideal for specifying hardware properties. An important use of PSL is as input to verification tools for *informal* dynamic verification (for example, simulation), as well as *formal* static verification (for example, property checkers).

Unlike the OVL set of monitors and the SystemVerilog assertion construct, which are used predominantly *during* RTL implementation, the PSL property language is suited for specifying architectural properties prior to *and* during RTL implementation. In addition, as a declarative form of specification, PSL is also suited for specifying interface properties during block-level refinement from an architectural model. Finally, the expressiveness of PSL makes it excellent for capturing implementation assertions and boundary assumptions during RTL implementation.

PSL is a large and expressive property language. However, while every PSL property can be checked in formal verification, not every PSL property can be checked in simulation. A subset of the PSL language lends itself to automatically generating simulation checkers (also referred to as monitors), which then can be used to check specific design properties during simulation. Hence, we are limiting our PSL introduction to a subset of the language that can be checked in both simulation *and* formal verification. A full description of the PSL language can be found in the Accellera PSL Formal Property Language Reference Manual [Accellera PSL-1.0 2003].

B.2 Operators and keywords

Table B.1 shows the PSL keywords, which are case-sensitive. This appendix discusses the keywords marked in **bold** type.

Table B.1 PSL keywords

A	E	next_a	U
AF	EF	next_a!	union
AG	EG	next_e	**until**
AX	EX	next_e!	**until!**
abort	**endpoint**	next_event	**until!_**
always	**eventually!**	next_event!	**until_**
and[1]		next_event_a!	
assert	F	next_event_e!	vmode
assume	fairness	**not**[1]	vprop
assume_guarantee	**fell**		vunit
	forall	**or**[1]	
before			W
before!	G	**property**	**whilenot**
before!_		**prev**	**whilenot!**
before_	in		**whilenot!_**
boolean	inf	restrict	**whilenot_**
	inherit	restrict_guarantee	**within**
clock	**is**[1]	**rose**	**within!**
const	**never**		**within!_**
cover	**next**	**sequence**	**within_**
	next!	strong	
default			X
		to[1]	X!

1. Keyword in VHDL flavor of PSL only

Table B.2 lists the operators available in PSL from highest to lowest precedence. This appendix addresses the operators in bold. Those in standard weight are beyond the scope of this appendix. Additional information on these operators can be found in the PSL Language Reference Manual [Accellera PSL-1.0 2003].

Table B.2 PSL operator precedence

<table>
<tr><td>HDL operators</td><td><code>operators from Verilog, VHDL, or EDL</code></td><td></td></tr>
<tr><td>Clocking operator</td><td><code>@</code></td><td>specifies clock expression while controls when the property is evaluated</td></tr>
<tr><td>SERE construction operators</td><td><code>;</code></td><td>temporal concatenation</td></tr>
<tr><td></td><td><code>[*]</code></td><td>consecutive repetition</td></tr>
<tr><td></td><td><code>[=]</code></td><td>non-consecutive repetition</td></tr>
<tr><td></td><td><code>[->]</code></td><td>goto repetition</td></tr>
<tr><td>Sequence implication operators</td><td><code>:</code></td><td>sequence fusion</td></tr>
<tr><td></td><td><code>|</code></td><td>sequence disjunction</td></tr>
<tr><td></td><td><code>&</code></td><td>non-length-matching sequence connection</td></tr>
<tr><td></td><td><code>&&</code></td><td>length-matching sequence connection</td></tr>
<tr><td>Foundation layer implication operators</td><td><code>|-></code></td><td>weak suffix implication</td></tr>
<tr><td></td><td><code>|-> !</code></td><td>strong suffix implication</td></tr>
<tr><td></td><td><code>|=></code></td><td>weak next suffix implication</td></tr>
<tr><td></td><td><code>|=> !</code></td><td>strong next suffix implication</td></tr>
<tr><td>Foundation layer occurrence operators</td><td><code>always</code></td><td>must hold, globally</td></tr>
<tr><td></td><td><code>never</code></td><td>must NOT hold, globally</td></tr>
<tr><td></td><td><code>eventually!</code></td><td>must hold at some time in the indefinite future</td></tr>
</table>

	next **next!** next_a next_a! next_e next_e! next_event next_event_a next_event_a! next_event_e next_event_e!	must hold at some specified future time or range of future times
	within **within!** **within!_** **within_**	must hold following completion of a sequence until a termination condition
	whilenot **whilenot!** **whilenot!_** **whilenot_**	must hold from the current cycle until a termination condition
Termination operators	**abort**	must hold, but future obligations may be canceled by a given event
	until **until!** **until!_** **until_**	must hold up to a given event
	before **before!** **before!_** **before_**	must hold at some time before a given event
LTL operators	X X! F G [U] [W]	same as next same as next! same as eventually! same as always same as until! same as until

PSL also defines a set of Options Branching Extension (OBE) operators, which are beyond the scope of this appendix. Please refer to the PSL LRM [Accellera PSL-1.0 2003] for details.

B.3 PSL Boolean layer

The *Boolean layer* of PSL provides for any Boolean expression valid within the language flavor of PSL being used (that is, Verilog or VHDL Boolean expressions). The result of the Boolean expression is a singular value of TRUE or FALSE. This is equivalent to a condition being evaluated within an `if` statement within Verilog or VHDL. Additionally, PSL provides a number of predefined functions that return Boolean values.

B.4 PSL Temporal Layer

The *temporal layer* is the heart of the PSL language; it describes properties of the design that have complex temporal relationships. Thus, unlike simple properties such as "signals a and b are mutually exclusive", the temporal layer allows PSL to describe relationships between signals, such as "*if signal c is asserted, then signal d must be asserted before signal e is asserted, but no more than 8 clock cycles later*".

Sequence Extended Regular Expressions (SERE)

PSL's *temporal layer* is based on an extension of regular expressions, called *Sequence Extended Regular Expressions* (SEREs), which is in many cases more concise and easier to read and write. The simplest SERE is a Boolean expression describing a Boolean event. More complicated SEREs are built from Boolean expressions using various SERE composition operators. A SERE is not a property on its own; it is a building block of a property; that is, properties are built from temporal operators applied to SEREs and Boolean expressions.

a sequence *holds*

Within this section, we refer to a sequence *holding*. This term indicates that the behavior described by the sequence or property is actually seen.

This section is organized by describing the composition operators first, followed by the temporal operators.

B.4.1 Named SERE

PSL allows us to name property definitions as shown in Syntax B-22. Note that DEF_SYM is '=' for Verilog and 'is' for VHDL.

Syntax B-1 Named property
Sequence_Declaration ::=   **sequence** Name [**(** Formal_Parameter_List **)**] DEF_SYM Sequence **;**

B.4.2 SERE concatenation (;) operator

The SERE concatenation operator (;) describes a sequence of events by specifying two sequences of events that must hold one after the other. That is, the second SERE starts one cycle after the first SERE completes.

The right operand of ';' is a SERE that is required to hold after the left operand completes. If either operand describes the empty sequence, then the concatenation holds if and only if the non-empty sequence holds.

Syntax B-2 Concatenation of sequences
SERE ::=   SERE **;** SERE

B.4.3 Consecutive repetition ([*]) operator

The SERE *consecutive repetition* operator ([*]) describes repeated consecutive concatenation of the same SERE.

Note the RANGE_SYM is ':' for Verilog and 'to' for VHDL.

Syntax B-3 SERE consecutive repetition

```
SERE ::=
        SERE [ * [ Count ] ]
       | [ * [ Count ] ]
       | SERE [ + ]
       | [ + ]

Count ::=
       Number | Range
Range ::=
       LowBound RANGE_SYM HighBound
LowBound ::=
       Number | MIN_VAL
HighBound ::=
       Number | MAX_VAL
```

The first operand of consecutive repetition is a SERE that is required to hold several consecutive times. The second operator is a number (or a range of numbers) that describes the number of times the SERE is repeated.

If the high value of the range is not specified (or is specified as inf), then the SERE must hold for at least the low value of the range. If the low value of the range is not specified (or is specified as '0') then the SERE must hold no more than the high value of the range times. If neither of the range values is defined, then the SERE is allowed to hold any number of times including zero, that is, the empty sequence is allowed.

When there is no SERE operand, and only a number (or a range), then the resulting SERE describes any sequence whose length is described by the second operand as above.

The notation '+' is a shortcut for a repetition of one or more times.

Syntax B-4 SERE consecutive repetition

```
SERE_A [ Number_n]              // SERE_A repeats exactly Number_n times

SERE_A [Number_n : Number_m]    // SERE_A repeats at least Number_n times
                                // no more than Number_m times

SERE_A [0 : Number_m]           // SERE_A is either empty or repeats no
                                // more than Number_m times

SERE_A [Number_n : inf]         // SERE_A repeats at least Number_n
                                // times

SERE_A [0 : inf]                // SERE_A is either empty or repeats some
                                // undefined number of times

SERE_A [+]                      // SERE_A evaluates one or more times

[* Number_n]                    // the sequence is of length Number_n

[* Number_n : Number_m]         // the length of the sequence is a number
                                // between Number_n and Number_m

[* 0 : Number_m]                // an empty sequence or a sequence of
                                // length Number_m at most

[* Number_n : inf]              // a sequence is of length Number_n at
                                // least

[* 0 : inf]                     // any sequence of events

[*]                             // any sequence of events (including the
                                // empty sequence)

[+]                             // any sequence of events of length one
                                // at least
```

B.4.4 Nonconsecutive repetition ([=]) operator

Nonconsecutive repetition allows for space between the repetition terms. The syntax for nonconsecutive repetition is the same as for consecutive repetition except the '*' operator is replaced with the '=' operator.

Note the `RANGE_SYM` is '`:`' for Verilog and '`to`' for VHDL.

Syntax B-5 SERE nonconsecutive repetition

```
SERE ::=
        Boolean [ = Count ]

Count ::=
        Number | Range
Range ::=
        LowBound RANGE_SYM HighBound
LowBound ::=
        Number | MIN_VAL
HighBound ::=
        Number | MAX_VAL
```

B.4.5 Goto repetition ([->]) operator

Goto repetition allows for space between the repetition of the terms. The repetition ends on the Boolean expression being found. This facilitates searching for a particular expression and then continuing the sequence at the point it is found.

Note the `RANGE_SYM` is '`:`' for Verilog and '`to`' for VHDL.

Syntax B-6 Goto repetition of a sequence

```
SERE ::=
        Boolean [ -> [ positive_Count ] ]

Count ::=
        Number | Range
Range ::=
        LowBound RANGE_SYM HighBound
LowBound ::=
        Number | MIN_VAL
HighBound ::=
        Number | MAX_VAL
```

B.4.6 Sequence fusion (:) operator

The *sequence fusion* operator specifies that two sequences overlap by one cycle. In this case, the second sequence starts the cycle that the first sequence ends.

Syntax B-7 Sequence fusion operator

```
SERE ::= SERE : SERE
```

B.4.7 Sequence non-length-matching (&) operator

The *sequence non-length-matching and* operator specifies that two sequences must hold and complete in different cycles (that is, they may be of different lengths).

Syntax B-8 Sequence non-length-matching and operator

```
SERE ::= SERE & SERE
```

B.4.8 Sequence length-matching (&&) operator

The *sequence length-matching and* operator specifies that two sequences must hold and complete in the same cycle (that is, they must be of the same length).

Syntax B-9 Sequence length-matching and operator

```
SERE ::= SERE && SERE
```

B.4.9 Sequence or (|) operator

The *sequence or* operator specifies that one of two alternative sequences must hold.

Syntax B-10 Sequence or operator

```
SERE ::= SERE | SERE
```

B.4.10 until* sequence operators

The until* operators (until, until!, until!_, and until_) specify that a property holds until a second property holds. The until* operators provide another way to move forward, this time while putting a requirement on the cycles in which we are moving.

Syntax B-11 until* operators

FL_Property ::=
    FL_Property **until!** FL_Property
  | FL_Property **until** FL_Property
  | FL_Property **until!_** FL_Property
  | FL_Property **until_** FL_Property

weak versus strong operators

The different flavors of this operator specify strong (until! and until!_) or weak (until and until_) operators. *Strong operators* require the terminating property to eventually occur, while *weak operators* do not. The inclusive operators (until_ and until!_) specify that the property must hold up to and including the cycle in which the terminating property holds, whereas the non-inclusive operators (until and until!) require the property to hold up to, but not necessarily including, the cycle in which the terminating property holds.

B.4.11 within* sequence operators

The within operators (within, within!, within!_, and within_) specify a window of time in which some other sequence should be seen. If the starting condition is a sequence, the completion of the sequence defines the beginning of the window.

Syntax B-12 within* operators

FL_Property ::=
    within! **(** Sequence_or_Boolean **,** Boolean **)** Sequence
  | **within** **(** Sequence_or_Boolean **,** Boolean **)** Sequence
  | **within!_** **(** Sequence_or_Boolean **,** Boolean **)** Sequence
  | **within_** **(** Sequence_or_Boolean **,** Boolean **)** Sequence

weak versus strong operators

The different flavors of this operator specify strong (within! and within!_) or weak (within and within_) operators. Strong operators require the ending condition to eventually occur, while weak operators do not. If the ending condition overlaps with the rightmost operand of the sequence, use the inclusive operators (within_ and within!_). The non-inclusive operators (within and within!) require that the rightmost operand of the sequence complete the cycle before the terminating condition.

B.4.12 next operator

The next operators allow us to be more specific about the timing; it takes us forward one clock cycle. The `next` operator comes in both weak (next) and strong (next!) forms. If the number parameter is present, it indicates the cycle at which the property on the right hand side holds.

For further information on the remaining family of next* operators, please refer to the PSL LRM.

Syntax B-13 next* operators

FL_Property ::=
  next FL_Property
  | **next!** FL_Property
  | **next [** Number **] (** FL_Property **)**
  | **next! [** Number **] (** FL_Property **)**

B.4.13 eventually! operator

While the next operator moves us forward exactly one cycle, the eventually! operator allows us to move forward without specifying exactly when to stop. This operator is a strong operator, which requires that the ending property or sequence actually occur.

Syntax B-14 eventually! operators

FL_Property ::=
  eventually! FL_Property
  | **eventually!** Sequence

B.4.14 before* operators

The before* operators provide an easy way to state that some signal must be asserted before some other signal.

Syntax B-15 before operator

FL_Property ::=
  FL_Property **before!** FL_Property
  | FL_Property **before** FL_Property
  | FL_Property **before!_** FL_Property
  | FL_Property **before_** FL_Property

weak versus strong operators

The different flavors of this operator specify strong (before! and before!_) or weak (before and before_) operators. Strong operators require the ending condition to eventually occur, while weak operators do not. If the ending condition overlaps with the rightmost operand of the sequence, use the inclusive operators (before_ and before!_). Use the non-inclusive operators (before and before!) to require the rightmost operand of the sequence to complete the cycle before the terminating condition.

B.4.15 abort operator

The abort operator provides a way to lift any future obligations of a property when some Boolean condition is observed.

Syntax B-16 abort operator

FL_Property ::=
  FL_Property **abort** Boolean

The `abort` operator is reminiscent of the `until` operator, but there is an important difference. Both "`f abort b`" and "`f until b`" specify that we should stop checking when `b` occurs. However, the abort operator removes future obligations of `f`, while the until operator does not.

B.4.16 Endpoint declaration

An endpoint for a sequence is defined in PSL with a named endpoint declaration. The name of an endpoint cannot be the same name as other named PSL declarations.

Syntax B-17 Endpoint declaration

Endpoint_Declaration ::=
    endpoint Name [**(** Formal_Parameter_List **)**] DEF_SYM Sequence **;**

B.4.17 Suffix implication operators

A SERE is not a PSL property in and of itself. In order to use a SERE to build a PSL property, we link a SERE with another PSL property or with another SERE using an implication operator. An implication operator can be read as "whenever we have a sequenceA, we expect to see sequenceB."

Syntax B-18 Suffix implication

FL_Property ::=
    Sequence **(** FL_Property **)**
  | Sequence **|->** Sequence [**!**]
  | Sequence **|=>** Sequence [**!**]

weak versus strong operators

The strong implication operators specify that the rightmost sequence must complete, whereas the weak implication operators do not. The suffix implication specifies that the rightmost sequence begins on the cycle in which the leftmost sequence ends. The suffix next implication specifies that the rightmost sequence begins on the cycle after the leftmost sequence ends.

B.4.18 Logical implication operator

The logical `if` implication operator specifies that if the leftmost property holds, then the rightmost property must hold.

Syntax B-19 Logical implication

FL_Property ::=
    FL_Property **->** FL_Property

B.4.19 always temporal operator

The always operator specifies one of the simplest temporal properties, which states that some Boolean expression must hold at all times.

Syntax B-20 always

FL_Property ::=
always FL_Property
| **always** Sequence

B.4.20 never temporal operator

The never operator allows us to specify an invariant property, which specifies conditions that must *never* hold.

Syntax B-21 never

FL_Property ::=
never FL_Property
| **never** Sequence

B.5 PSL properties

B.5.1 Property declaration

The building blocks of Boolean expressions and sequences described in previous sections create PSL properties. Properties capture the temporal relationships between these building blocks. Properties are grouped using parentheses (( )).

B.5.2 Named properties

PSL allows us to name property definitions as shown in Syntax B-22. Note that DEF_SYM is '`=`' for Verilog and '`is`' for VHDL.

Syntax B-22 Named property

Property_Declaration ::=
  property Name [**(** Formal_Parameter_List **)**] DEF_SYM Property **;**

B.5.3 Property clocking

PSL allows us to declare a default clock, explicitly declare a clock associated with a property, or declare that "clock cycle" and "next point in time" are equivalent. A clock expression is any Boolean expression. A PSL property may refer to multiple clocks.

Syntax B-23 Default clock declaration

Clock_Declaration ::=
  default clock DEF_SYM Boolean **;**(see B.8.3.7)

Syntax B-24 Clocked property or SERE

SERE ::=
  SERE **@** *clock*_Boolean

FL_Property ::=
  FL_Property **@** *clock*_Boolean [**!**]

B.5.4 forall property replication

PSL provides an easy way to replicate properties that are the same except for specific parameters. Example B-1 shows the syntax for the `forall` operator.

Example B-1 forall property replicatation syntax

Property ::=
    forall Name [IndexRange] **in** ValueSet **:** Property

B.6 The verification layer

The *verification layer* tells the verification tools what to do with the properties described by the temporal layer. For example, the Verification layer contains directives that tell a tool to assert a property (that is, to verify that it holds) or to check that a specified sequence is covered by some test case.

B.6.1 assert directive

The assert directive verifies that a property holds. If the property does not hold, an error is raised.

Syntax B-25 assert directive

Assert_Statement ::=
    assert Property **;**(see B.8.3.4)

B.6.2 assume directive

The assume directive defines constraints to guide a verification tool.

Syntax B-26 assume directive

Assume_Statement ::=
    assume Property **;**(see B.8.3.4)

B.6.3 cover directive

The cover directive instructs the tool to indicate if a property has been exercised by the test suite or given constraints.

Syntax B-27 cover directive

Cover_Statement ::=
    cover Sequence **;**(see B.8.3.5)

B.7 The modeling layer

The *modeling layer* models the behavior of design inputs (for tools such as formal verification tools that do not use test cases), and auxiliary hardware that is not part of the design but is needed for verification. This layer is, for the most part, outside the scope of this appendix. However, in this section we discuss a few useful functions that are defined by the modeling layer. These built-in functions are only defined for the Verilog flavor of PSL.

B.7.1 rose() and fell() functions

PSL provides the rose() and fell() functions, which are similar to `posedge` and `negedge` in Verilog. They take a Boolean signal and result in a Boolean value that is true if the value of the argument is the inverse of the previous cycle. For instance, rose() returns true when the current value is 1 and the value was 0 on the previous cycle. Similarly, fell() returns true when the current value is 0 and the value was 1 on the previous cycle.

Syntax B-28 rose() and fall()

Built_In_Function_Call ::=
    rose (Boolean **)**
    | **fell (** Boolean **)**

B.7.2 prev() and next() functions

The prev() and next() functions return the value of its argument on the previous or next cycle, respectively. An optional second argument specifies the number of cycles before or after the current cycle from which to return the value.

Syntax B-29 prev() and next()

Built_In_Function_Call ::=
    prev (HDL_or_PSL_Expression [**,** Number] **)**
    | **next (** Boolean **)**

B.8 BNF

The PSL BNF represented here is the property specification subset.

B.8.1 Verilog Extensions

For the Verilog flavor, PSL extends the forms of declaration that can be used in the modeling layer by defining two additional forms of type declaration. PSL also adds an additional form of expression for both Verilog and VHDL flavors.

Extended_Verilog_Declaration ::=
    *Verilog*_module_or_generate_item_declaration
    | Extended_Verilog_Type_Declaration

Extended_Verilog_Type_Declaration ::=
    integer Integer_Range list_of_variable_identifiers **;**
    | **struct {** Declaration_List **}** list_of_variable_identifiers **;**

Integer_Range ::=
    (constant_expression **:** constant_expression **)**

Declaration_List ::=
    HDL_Variable_or_Net_Declaration {
    HDL_Variable_or_Net_Declaration }

HDL_Variable_or_Net_Declaration ::=
    net_declaration
    | reg_declaration
    | integer_declaration

Extended_Verilog_Expression ::=
*Verilog*_expression
| *Verilog*_Union_Expression

Extended_VHDL_Expression ::=
*VHDL*_expression
| *VHDL*_Union_Expression

Union_Expression ::=
HDL_or_PSL_Expression **union**
HDL_or_PSL_Expression

B.8.2 Flavor macros

Flavor Macro PATH_SYM =
Verilog: **.** / VHDL: **:** / EDL: **/**
Flavor Macro HDL_ID =
Verilog: *Verilog*_Identifier / VHDL: *VHDL*_Identifier /
EDL: *EDL*_Identifier
Flavor Macro DEF_SYM =
Verilog: **=** / VHDL: **is** / EDL: **:=**
Flavor Macro RANGE_SYM =
Verilog: **:** / VHDL: **to** / EDL: **..**
Flavor Macro AND_OP =
Verilog: **&&** / VHDL: **and** / EDL: **&**
Flavor Macro OR_OP =
Verilog: **||** / VHDL: **or** / EDL: **|**
Flavor Macro NOT_OP =
Verilog: **!** / VHDL: **not** / EDL: **!**
Flavor Macro MIN_VAL =
Verilog: **0** / VHDL: **0** / EDL: *null*
Flavor Macro MAX_VAL =
Verilog: **inf** / VHDL: **inf** / EDL: *null*
Flavor Macro HDL_EXPR =
Verilog: Extended_Verilog_Expression / VHDL:
Extended_VHDL_Expression
/ EDL: *EDL*_Expression
Flavor Macro HDL_UNIT =
Verilog: *Verilog*_module_declaration / VHDL:
*VHDL*_design_unit / EDL: *EDL*_module_declaration
Flavor Macro HDL_DECL =
Verilog: Extended_Verilog_Declaration / VHDL:
*VHDL*_declaration
/ EDL: *EDL*_module_item_declaration
Flavor Macro HDL_STMT =

Verilog: *Verilog*_module_or_generate_item / VHDL:
*VHDL*_concurrent_statement
/ EDL: *EDL*_module_item

Flavor Macro LEFT_SYM =
Verilog: **[** / VHDL: **(** / EDL: **(**

Flavor Macro RIGHT_SYM =
Verilog: **]** / VHDL: **)** / EDL: **)**

B.8.3 Syntax productions

B.8.3.1 Verification units

PSL_Specification ::=
{ Verification_Item }

Verification_Item ::=
HDL_UNIT | Verification_Unit

Verification_Unit ::=
VUnitType Name [**(** Hierarchical_HDL_Name **)**] **{**
{ Inherit_Spec }
{ VUnit_Item }
}

VUnitType ::=
vunit | **vprop** | **vmode**

Name ::=
HDL_ID

Hierarchical_HDL_Name ::=
*module*_Name { PATH_SYM *instance*_Name }

Inherit_Spec ::=
inherit *vunit*_Name { **,** *vunit*_Name } **;**

VUnit_Item ::=
HDL_Decl_or_Stmt
| PSL_Declaration (see B.8.3.2)
| Verification_Directive(see B.8.3.3)

HDL_Decl_or_Stmt ::=
HDL_DECL | HDL_STMT

B.8.3.2 PSL declarations

PSL_Declaration ::=
Property_Declaration
| Sequence_Declaration
| Endpoint_Declaration
| Clock_Declaration

Property_Declaration ::=
property Name [**(** Formal_Parameter_List **)**]
DEF_SYM Property **;**

Formal_Parameter_List ::=
Formal_Parameter { **;** Formal_Parameter }

Formal_Parameter ::=
ParamKind Name { **,** Name }

ParamKind ::=
const | **boolean** | **property** | **sequence**

Sequence_Declaration ::=
sequence Name [**(** Formal_Parameter_List **)**]
DEF_SYM Sequence **;**(see B.8.3.5)

Endpoint_Declaration ::=
endpoint Name [**(** Formal_Parameter_List **)**]
DEF_SYM Sequence **;**(see B.8.3.5)

Clock_Declaration ::=
default clock DEF_SYM Boolean **;**(see B.8.3.7)

Actual_Parameter_List ::=
Actual_Parameter { **,** Actual_Parameter }

Actual_Parameter ::=
Number | Boolean | Property | Sequence (see B.8.3.7) (see B.8.3.7) (see B.8.3.4) (see B.8.3.5)

B.8.3.3 PSL statements

Verification_Directive ::=
Assert_Statement
| Assume_Statement
| Assume_Guarantee_Statement
| Restrict_Statement
| Restrict_Guarantee_Statement
| Cover_Statement
| Fairness_Statement

Assert_Statement ::=
assert Property **;**(see B.8.3.4)

Assume_Statement ::=
assume Property **;**(see B.8.3.4)

Assume_Guarantee_Statement ::=
assume_guarantee Property **;**(see B.8.3.4)

Restrict_Statement ::=
restrict Sequence **;**(see B.8.3.5)

Restrict_Guarantee_Statement ::=
restrict_guarantee Sequence **;**(see B.8.3.5)

Cover_Statement ::=

cover Sequence **;**(see B.8.3.5)

Fairness_Statement ::=
fairness Boolean **;**
| **strong fairness** Boolean **,** Boolean **;**(see B.8.3.7)

B.8.3.4 PSL properties

Property ::=
Replicator Property
| FL_Property
| OBE_Property

Replicator ::=
forall Name [IndexRange] **in** ValueSet **:**

IndexRange ::=
LEFT_SYM *finite*_Range RIGHT_SYM

ValueSet ::=
{ ValueRange { **,** ValueRange } **}**
| **boolean**

ValueRange ::=
Value (see B.8.3.7)
| *finite*_Range(see B.8.3.5)

FL_Property ::=
Boolean(see B.8.3.7)
| **(** FL_Property **)**
| *property*_Name [**(** Actual_Parameter_List **)**]

| FL_Property **@** *clock*_Boolean [**!**]
| FL_Property **abort** Boolean

: Logical Operators :
| NOT_OP FL_Property
| FL_Property AND_OP FL_Property
| FL_Property OR_OP FL_Property
:
| FL_Property **->** FL_Property
| FL_Property **<->** FL_Property

: Primitive LTL Operators :
| **X** FL_Property
| **X!** FL_Property
| **F** FL_Property
| **G** FL_Property
| **[** FL_Property **U** FL_Property **]**
| **[** FL_Property **W** FL_Property **]**

: Simple Temporal Operators :
| **always** FL_Property

| **never** FL_Property
| **next** FL_Property
| **next!** FL_Property
| **eventually!** FL_Property
:
| FL_Property **until!** FL_Property
| FL_Property **until** FL_Property
| FL_Property **until!_** FL_Property
| FL_Property **until_** FL_Property
:
| FL_Property **before!** FL_Property
| FL_Property **before** FL_Property
| FL_Property **before!_** FL_Property
| FL_Property **before_** FL_Property
: Extended Next (Event) Operators :(see B.8.3.7)
| **X [** Number **] (** FL_Property **)**
| **X! [** Number **] (** FL_Property **)**
| **next [** Number **] (** FL_Property **)**
| **next! [** Number **] (** FL_Property **)**
:(see B.8.3.5)
| **next_a [** *finite*_Range **] (** FL_Property **)**
| **next_a! [** *finite*_Range **] (** FL_Property **)**
| **next_e [** *finite*_Range **] (** FL_Property **)**
| **next_e! [** *finite*_Range **] (** FL_Property **)**
:
| **next_event! (** Boolean **) (** FL_Property **)**
| **next_event (** Boolean **) (** FL_Property **)**
| **next_event! (** Boolean **) [** *positive*_Number **] (** FL_Property **)**
| **next_event (** Boolean **) [** *positive*_Number **] (** FL_Property **)**
:
| **next_event_a! (** Boolean **) [** *finite_positive*_Range **] (** FL_Property **)**
| **next_event_a (** Boolean **) [** *finite_positive*_Range **] (** FL_Property **)**
| **next_event_e! (** Boolean **) [** *finite_positive*_Range **] (** FL_Property **)**
| **next_event_e (** Boolean **) [** *finite_positive*_Range **] (** FL_Property **)**
: Operators on SEREs :(see B.8.3.5)
| Sequence **(** FL_Property **)**
| Sequence **|->** Sequence [**!**]
| Sequence **|=>** Sequence [**!**]

:
| **always** Sequence
| **never** Sequence
| **eventually!** Sequence
:
| **within!** **(** Sequence_or_Boolean **,** Boolean **)** Sequence
| **within** **(** Sequence_or_Boolean **,** Boolean **)** Sequence
| **within!_** **(** Sequence_or_Boolean **,** Boolean **)**
Sequence
| **within_** **(** Sequence_or_Boolean **,** Boolean **)** Sequence
:
| **whilenot!** **(** Boolean **)** Sequence
| **whilenot** **(** Boolean **)** Sequence
| **whilenot!_** **(** Boolean **)** Sequence
| **whilenot_** **(** Boolean **)** Sequence

Sequence_or_Boolean ::=
Sequence | Boolean

B.8.3.5 Sequences

Sequence ::=
{ SERE **}**
| *sequence*_Name [**(** Actual_Parameter_List **)**]

B.8.3.6 Sugar extended regular expressions

SERE ::=
Boolean(see B.8.3.7)
| Sequence
| SERE **@** *clock*_Boolean
: Composition Operators :
| SERE **;** SERE
| Sequence **:** Sequence
| Sequence AndOrOp Sequence
: RegExp Qualifiers :
| SERE **[** ***** [Count] **]**
| **[** ***** [Count] **]**
| SERE **[+]**
| **[+]**
:
| Boolean **[=** Count **]**
| Boolean **[->** [*positive*_Count] **]**
AndOrOp ::=

&& | **&** | |

Count ::=
  Number | Range

Range ::=
  LowBound RANGE_SYM HighBound

LowBound ::=
  Number | MIN_VAL

HighBound ::=
  Number | MAX_VAL

B.8.3.7 Forms of expression

Value ::=
  Boolean | Number

Boolean ::=
  *boolean*_HDL_or_PSL_Expression

HDL_or_PSL_Expression ::=
  HDL_Expression
  | *endpoint*_Name [**(** Actual_Parameter_List **)**]
  | Built_In_Function_Call
  | HDL_or_PSL_Expression **union**
  HDL_or_PSL_Expression

HDL_Expression ::=
  HDL_EXPR

Built_In_Function_Call ::=
  rose **(** Boolean **)**
  | **fell** **(** Boolean **)**
  | **prev** **(** HDL_or_PSL_Expression [**,** Number] **)**
  | **next** **(** Boolean **)**

Number ::=
  *integer*_HDL_Expression

B.8.3.8 Optional branching extension

OBE_Property ::=
  Boolean
  | **(** OBE_Property **)**
  | *property*_Name [**(** Actual_Parameter_List **)**]

: Logical Operators :
  | **!** OBE_Property
  | OBE_Property **&** OBE_Property
  | OBE_Property **|** OBE_Property
  | OBE_Property **->** OBE_Property

| OBE_Property **<->** OBE_Property

: Universal Operators :

| **AX** OBE_Property

| **AG** OBE_Property

| **AF** OBE_Property

| **A [** OBE_Property **U** OBE_Property **]**

: Existential Operators :

| **EX** OBE_Property

| **EG** OBE_Property

| **EF** OBE_Property

| **E [** OBE_Property **U** OBE_Property **]**

APPENDIX

C

SYSTEMVERILOG ASSERTIONS

C.1 . Introduction to SystemVerilog

SystemVerilog is incorporating the concepts discussed in Chapter 3, "Specifying RTL Properties" on page 57. See the Accellera SystemVerilog 3.1 specification for the full standard of the language.

The BNF described here follows a few conventions.

- Keywords are in bold
- Required syntax characters are marked with single quotes

Production names not found in this text are part of the remainder of SystemVerilog 3.1 BNF.

C.2 Operator and keywords

SystemVerilog is introducing the following operators and keywords. These directly relate to sequences and properties. The SystemVerilog operators for data types are available for relating Boolean and vector expressions within sequence and property definitions.

Keyword Table

Keywords specific to assertions sequences, properties and templates.			
`assert`	`cover`	`property`	`endproperty`

Keywords specific to assertions sequences, properties and templates.			
or	intersect	within	throughout
and	first_match	ended	matched
template	endtemplate	sequence	endsequence

Operator table

Name	Operator	Description
Consecutive repetition	s1 [* N:M]	Repetition of s1 N, or between N to M times.
Goto repetition	s1 [*-> N:M]	Repetition of s1 (N to M times) in nonconsecutive cycles, ending on s1.
Nonconsecutive repetition	s1 [*= N:M]	Repetition of s1 N to M times in nonconsecutive cycles, maybe ending on s1.
Temporal delay	## N ## [N:M]	Concatenation of two sequence elements.
And	s1 & s2	Both sequences occur at some time.
Intersection	s1 **intersect** s2	Both sequences occur at the same time.
Or	s1 **or** s2	Either sequence occurs.
Boolean until	b **throughout** s1	B must be true until sequence s1 completes (results in a match).
Within	s1 **within** s2	s1 and s2 must occur. Lengths of s1 and s2 must follow s1 <= s2.
Ended	s1.**ended**	Sequence s1 matched (ended) at this time.
Matched (from different clock domain)	s1.**matched**	Sequence s1 (on another clock) ended at this time.
First match	**first_match**(s1)	First occurrence of sequence, rest ignored.
Overlapping implication	s1 |-> s2	If s1 occurs, s2 (starting this cycle) must occur, else true.
Non-overlapping implication	s1 |=> s2	If s1 occurs, s2 (starting next cycle) must occur, else true.

SystemFunction table

$rose	$fell	$stable

$countones	$onehot	$onehot0
$inset	$insetz	$isunknown
$past		

Operator precedence table

Repetition(consecutive, nonconsecutive, goto)	
And	Intersection
Or	
Boolean until (Throughout)	
Within	
Delay (##)	
Implication	

C.3 Sequence and property operations

The new sequence operators defined for SystemVerilog allow us to compose expressions into temporal sequences. These sequences are the building blocks of properties and concurrent assertions. The first four allow us to construct sequences, while the remaining operators allow us to perform operations on a sequence as well as compose a complex sequence from simpler sequences.

C.3.1 Temporal delay

The temporal delay operator "##" constructs larger sequences by combining small sequences.

Example C-1 Construction of sequences with temporal delay

```
sequence_expr ::= sequence_expr '##'  cycle_delay_range
                    sequence_expr

cycle_delay_range ::=
      constant_range_expression
    | '[' constant_range_expression ':' constant_range_expression ']'
    | '['constant_range_expression ':' '$'   ']' -- Infinite range.
```

A temporal delay may begin a sequence. The range may be a single constant amount or a range of time. All times may be used to match the following sequence. The range is interpreted as follows:

- ##0 a - same as (a)
- ##1 a - same as (1 ##1 a)
- ##[0:1] a - same as a or (1 ##1 a)

When the symbol **$** is used, the range is infinite.

The right side sequence of a concatenation is expected on the following cycle unless the delay has the value 0. Each subsequent element of a concatenation further advances time one cycle. Examples:

- 1 ##1 a - means a must be true in cycle 1.
- 1 ##1 a ##1 b - means a must be true in cycle 1, b must be true in cycle 2.
- a ##[0:2] b same as a & b or (a ##1 b) or (a ##1 1 ##1 b)
- a ##[2:3] b same as (a ##1 1 ##1 b) or (a ##1 1 ##1 1 ##1 b)

Note:
When a range is used, it is possible for the sequence to match with each value within the range. Thus, we must take into account that there may be multiple possible matches when we write sequences with a range.

C.3.2 Consecutive repetition

consecutive repetition of a sequence applies to a single element or a sequence.

Example C-2 Consecutive repetition of a sequence

```
sequence_expr ::= expression { ',' function_blocking_assign }
                             '[' * const_range_expression ']'
    | '(' sequence_expr ')' '[' '*' const_range_expression ']'

const_range_expression ::=
      constant_range_expression
    | constant_range_expression ':' constant_range_expression
    | constant_range_expression ':' '$'

function_blocking_assign ::= variable '=' expression
```

When using a repetition, the element or sequence must occur starting in the next cycle for each repetition expected. Examples are:

- a [* 0] ##1 b same as (b)
- a [* 2] ##1 b same as (a ##1 a ##1 b)
- a [* 1:2] ##1 bsame as (a ##1 b) or (a ##1 a ##1 b)
- (a ##1 b) [* 2] same as (a ##1 b ##1 a ##1 b)

Note:

When a range is used, it is possible for the sequence to match with each value within the range. Thus, we must take into account that there may be multiple possible matches when we write sequences with a range.

C.3.3 Goto repetition

goto repetition allows for space between the repetition of the terms. The repetition ends on the Boolean expression being found. This facilitates searching for a particular expression and then continuing the sequence at the point it is found.

Example C-3 Goto repetition of a sequence

```
sequence_expr ::=    expression '[' '*->' const_range_expression ']'
 const_range_expression ::=constant_expression
     | constant_expression ':' constant_expression
     | constant_expression ':' '$'
```

goto repetition is defined in terms of the other operators as:

```
s [*-> min:max] ::= (!s[*0:$] ##1 s) [* min:max]
```

Examples are:

- a [*->0] ##1 b same as (b)
- a [*->1] ##1 b same as (!a [*0:$] ##1 a ##1 b)
- a [*->2] ##1 b same as (!a [*0:$] ##1 a ##1 !a [*0:$] ##1 a ##1 b)

Note:

When a range is used, it is possible for the sequence to match with each value within the range. Thus, we must take into account that there may be multiple possible matches when we write sequences with a range.

C.3.4 Nonconsecutive repetition

nonconsecutive repetition, like goto repetition allows for space between the repetition of the expression. At the end of the repetition, however, there can be additional space after the repeated expression.

Example C-4 Nonconsecutive repetition of a sequence

```
  sequence_expr ::= expression '[' '*=' const_range_expression ']'
const_range_expression ::=  constant_range_expression
    | constant_range_expression ':' constant_range_expression
    | constant_range_expression ':' '$'
```

nonconsecutive repetition is defined in terms of the other operators as:

```
s [*= min:max] ::= (!s[*0:$] ##1 s ##1 !s[*0:$])
                   [* min:max ]
```

Examples are:

- a [*=0] ##1 b same as (b)
- a [*=1] ##1 b same as (!a [*0:$] ##1 a ##1 !a [*0:$] ##1 b)
- a [*=2] ##1 b same as (!a [*0:$] ##1 a ##1 !a [*0:$] ##1 a ##1 !a [*0:$] ##1 b)

Note:

When a range is used, it is possible for the sequence to match with each value within the range. Thus, we must take into account that there may be multiple possible matches when we write sequences with a range.

C.3.5 Sequence and

A sequence and produces a match once both sequences produce a match (the end point may not match). A match occurs until the endpoint of the longer sequence (provided the shorter sequence produces one match).

Example C-5 Sequence and (non-matching length)

```
sequence_expr ::= sequence_expr and sequence_expr
```

This operation is defined in terms of the next operator (intersect) as:

```
s and t ::= ( (s ##1 1 [*0:$]) intersect t)
            or (s intersect (t ##1 1 [*0:$])).
```

Examples of sequence and'ing are (() mean no match):

- (a ##1 b) **and** () same as ()
- (a ##1 b) **and** (c ##1 d) same as (a & c ##1 b & d)
- (a ##[1:2] b) **and** (c ##[3] d)same as(a & c ##1 b ##1 1 ##1 d)
  or (a & c ##1 1 ##1 b ##1 d)

Note:

It is possible for this operation to match with each value of a range or repeat in the operands. Thus, we must take into account that there may be multiple possible matches when we use this operator.

C.3.6 Sequence intersection

An intersection of two sequences is like an and of two sequences (both sequences produce a match). This operator also requires the length of the sequences to match. That is, the match point of both sequences must be the same time. With multiple matches of each sequencc, a match occurs each time both sequences produce a match.

Example C-6 Sequence and (matching length)

```
sequence_expr ::= sequence_expr 'intersect sequence_expr
```

Examples of a sequence intersection are (() means no match):

- (1) **intersect** () same as ()
- ##1 a **intersect** ##2] b same as ()
- ##2 a **intersect** ##2] b match if (##2 (a & b))
- ##[1:2] a **intersect** ##[2:3] b match if (1 ##1 a&b)
  or (1 ##1 a&b ##1 b)

- ##[1:3] a **intersect** ## [1:3] bmatch if (##1 a&b)
or (##2 a&b)
or (##3 a&b)

Note:

It is possible for this operation to match with each value of a range or repeat in the operands. Thus, we must take into account that there may be multiple possible matches when we use this operator.

C.3.7 Sequence or

A match on either sequence results in a match for this operation.

Example C-7 Sequence or

```
sequence_expr ::= sequence_expr or sequence_expr
```

Examples of sequence or are:

- () **or** () same as ()
- (## 2 a **or** ## [1:2] b) match if
(b) or (##1 b) or (## 2 a) or (##2 b)

Note:

It is possible for this operation to match with each value of a range or repeat in the operands. Thus, we must take into account that there may be multiple possible matches when we use this operator.

C.3.8 Boolean until (throughout)

This operator matches a Boolean value throughout a sequence. The operator produces a match if the sequence matches and the Boolean expression is true until the end of the sequence.

Example C-8 (Boolean) throughout sequence

```
sequence_expr ::= expression throughout sequence_expr
```

The throughout operator is defined in terms of the intersect sequence operators. Its definition is:

```
(b throughout s) ::= ( b [*0:$] intersect s)
```

Examples include:

- 0 **throughout** (1)   same as()
- 1 **throughout** ##1 a   same as##1 a
- a **throughout** ##2 b   same as(a ##1 a & b)
- a **throughout** (b ##[1:3] c) same as (a&b ##1 a ##1 a &c)
  or (a&b ##1 a ##1 a ##1 a &c)
  or (a&b ##1 a ##1 a ##1 a ##1 a &c)

Note:

It is possible for this operation to match with each value of a range or repeat in the operands. Thus, we must take into account that there may be multiple possible matches when we use this operator.

C.3.9 Within sequence

The within operator determines if one sequence matches within the length of another sequence.

Example C-9 Within sequence

```
sequence_expr ::= sequence_expr within sequence_expr
```

The within operator is defined in terms of the intersect sequence operators. Its definition is:

```
s1 within s2 ::= ((1 *[*0:$] ##1 s1 ##1
        1 [*0:$]) intersect s2)
```

Examples are (() means no match):

- () **within** (1)   same as   ()
- (1) **within** ()   same as   ()
- (a) **within** ## [1:2] b   same as (a&b) or (b ##1 a)
  or (a ##1 b) or (##1 a&b)

Note:

It is possible for this operation to match with each value of a range or repeat in the operands. Thus, we must take into account that there may be multiple possible matches when we use this operator.

C.3.10 Ended

The ended method checks for the end of a matching sequence. This is in contrast to using a matching sequence from the beginning timepoint, which is obtained when we use only the sequence name.

Example C-10 Ended sequence

```
boolean_expr_op ::= sequence_identifier. 'ended'
sequence_identifier ::= (identifier defined as a sequence)
```

Examples include:

```
sequence a1;
   @(posedge clk) (c ##1 b ##1 d);
endsequence
```

- `(a ##1  a1.ended)` same as `(c ##1 b & a ##1 d)`
- `(a ##2  a1.ended)` same as `(c &a ## b ##1 d)`

Note the position of '`a`' relative to the end of sequence '`a1`' (term 'd'). Compare this with the following sequence where 'a' occurs before '`a1`'

- `(a ##1  a1)` same as `(a ##1 c ##1 b ##1 d)`

C.3.11 Matched

The matched method operates similarly to the ended method; however, this method is used when the sequence of the method call uses a different clock than the sequence being called.

Example C-11 Matched sequence

```
boolean_expr_op ::= sequence_identifier '.' matched

sequence_identifier ::= (identifier of type sequence)
```

C.3.12 First match

The first match operator returns only the first match from a sequence. The remaining are suppressed.

Example C-12 First match

```
sequence_expr ::= first_match sequence_expr
```

Examples of first_match are:

- **first_match** (1 [*1:5]) same as (1)
- **first_match** (##[0:4] 1)same as (1)
- **first_match** (##[0:1] a) same as (a) or (!a ##1 a)
- **first_match** (s1 **intersect** s2) same as (s1 **intersect first_match**(s2)
- **first_match** (b **throughout** s1) same as (b **throughout first_match**(s1))
- **first_match**(s1 **within** s2) same as (s1 **within first_match** (s2))

Note:

Use of a range on the delay operators or a range on the repetition operators can cause multiple matches. Use of first_match can be helpful to suppress the subsequent matches. These additional matches may cause a false firing that is solved with this operator.

C.3.13 Implication

As a convenience, there are two forms of implication, overlapping and non-overlapping. The overlap occurs between the final cycle of the left-hand side and the starting cycle of the right-hand side operands. For the overlapping form, the right-hand side starts on the current cycle (that the left-hand side matched). The non-overlapping form has the right-hand side start the subsequent cycle. implication is similar in concept to an if() statement.

implication uses the left side operand as a test condition to determine whether the right side operand should be evaluated or the operation should just return a true result.

Example C-13 Overlapping implication

```
prop_expr ::= sequence_expr '|->' sequence_expr
```

Example C-14 Non-overlapping (suffix) Implication

```
prop_expr ::= sequence_expr '|=>' sequence_expr
```

Examples include:

- a |-> b same as a ? b : 1'b1
- (a ##1 b) |-> (c) same as (a ##1 b) ? c : 1'b1
- (a ##1 b) |=> (c ##1 d) same as (a ##1 b) |-> ##1 c ##1 d)

C.4 Sequences and properties

SystemVerilog allows for declarations of both sequences and properties. A property differs from a sequence in that it contains a clock specification for the entire property, a term to disable property evaluations and a keyword to invert the result of a property (match or fail). Properties also allow implication operations.

Example C-15 Property declaration

```
  property_declaration ::=
      property property_identifier [ property_formal_list ] ';'
          { property_decl_item }
          property_spec ';'
        endproperty [ ':' property_identifier ]

property_formal_list ::= '(' formal_list_item { , formal_list_item } ')'
formal_list_item ::=  formal_identifier [ '=' actual_arg_expr ]
actual_arg_expr ::=  expression | identifier |  event_control

property_decl_item ::=  sequence_declaration
      | list_of_variable_identifiers_or_assignments

property_spec ::= [ event_control ] property_modifier property_expr
      | property_expr
property_modifier ::= [ disable iff '(' expression ')' ] not
      | disable iff  '('expression ')'  [ not ]
property_expression ::=  sequence_spec | property_implication

property_implication ::=
        sequence_expr              |->  [ not ] sequence_expr
      | clocked_sequence         |-> [ not ] sequence_expr
      | sequence_expr              |=> [ not ] sequence_expr
      | clocked_sequence         |=> [ not ] sequence_expr
      | multi_clock_sequence   |=> [ not ] multi_clock_sequence

multi_clock_sequence  ::= clocked_sequence { '##' clocked_sequence }
clocked_sequence ::= event_control sequence_expr
event_control ::= '@' event_identifier | '@' (event_expression ')'
```

Properties can be complete definitions useful with other properties, or they can be used for verification as an assertion or as a coverage point. Properties can also contain parameters to be specified when they are used in these other contexts. Examples include:

```
property req_t1_start;
        @(posedge clk)  req && req_tag == t1;
endproperty
property illegal_op;
        @(posedge clk)  not req && cmd == 4;
endproperty
property starts_at(start, grant);
        @(posedge clk) (grant ##1 grant & start);
endproperty
```

Sequences are also declared. They use syntax similar to properties.

Example C-16 Sequence declaration

```
sequence_declaration ::=
     sequence sequence_identifier [sequence_formal_list ] ';'
       { sequence_decl_item }
       sequence_spec ';'
     endsequence [ ':' sequence_identifier ]
sequence_formal_list ::=
      '(' formal_list_item { ',' formal_list_item } ')'
sequence_decl_item ::=
      list_of_variable_identifiers_or_assignments.
sequence_spec ::=
      multi_clock_sequence | sequence_expr
```

They can be defined within properties and as separate elements. They also can be declared with parameters that are specified when used in other contexts. These elements, coupled with the following directives, allow one to define and utilize properties to follow an assertion-based design methodology. Examples include:

```
sequence op_retry;
   (req ##1 retry);
endsequence
sequence cache_fill(req, done, fill);
   (req ##1 done [*-> 1] ##1 fill [*-> 1]);
endsequence
```

Sequence specifications can use all the operators defined above, except implications. Example C-17 shows the BNF.

Example C-17 Sequence specification

```
cycle_delay_range ::=
       '##' constant_expression
    |  '##' '[' const_range_expression ']'
const_range_expression ::=
       constant_expression : constant_expression
    |  constant_expression : '$'

sequence_expr ::=
    [ cycle_delay_range ] sequence_expr
    { cycle_delay_range   sequence_expr }
    | expression { ',' function_blocking_assign } [ boolean_repeat ]
    | first_match '(' sequence_expr ')'
    | expression throughout sequence_expr
    | sequence_instance [ consecutive_repeat_operator ]
    | '(' sequence_expr ')'  [ consecutive_repeat_operator ]
    | sequence_expr and sequence_expr
    | sequence_expr intersect sequence_expr
    | sequence_expr or sequence_expr
    | sequence_expr within sequence_expr
    |  boolean_expr_op

boolean_repeat ::= consecutive_repeat_operator
    |  nonconsecutive_repeat_operator
    |  goto_repeat_operator

consecutive_repeat_operator     ::=
     '[' '*'   const_range_expression ']'

nonconsecutive_repeat_operator ::=
     '['  '*=' const_range_expression ']'

goto_repeat_operator ::=
     '['  '*->' const_range_expression ']'

boolean_expr_op ::=
     sequence_identifier '.' ended
   | sequence_identifier '.' matched
   | $countones '(' expression ')'
   | value_change_functions

value_change_functions  ::=
     $rose '(' expression ')'
   | $past    '(' expression [ ',' number_of_ticks] ')'
   | $fell     '(' expression ')'
   | $stable '('expression ')'
```

C.5 Assert and cover statements.

Property directives define how to use properties (and sequences) for specific works. These statements utilize all the elements above to define how they are to be used for a given design.

Example C-18 Property directives

```
immediate_assert_statement ::=
    assert ( expression ) action_block

concurrent_assert_statement ::=
    assert property '(' property_instance ')' action_block
  | assert property '(' sequence_instance ')' action_block
  | assert property '(' property_spec ')' action_block

concurrent_cover_statement ::=
    cover property '(' property_instance ')' statement_or_null
  | cover property '(' sequence_instance ')' statement_or_null
  | cover property '(' property_spec ')' statement_or_null

action_block ::=
    statement    [ else statement_or_null ]
  | [statement_or_null] else statement_or_null

statement_or_null ::=
    statement    |   ';'
```

As an assertion, properties are evaluated, and when they do not match the desirable sequence, they fail and produce an error message by default (or they execute the else statement set). When they match a sequence, the first (pass) statement set is executed (like an **if()** statement).

As a cover directive, properties are evaluated, and when they succeed, they execute the first (pass) statement set. If they fail to match a sequence they execute the else statement set, if any. Examples include:

- **assert property** (illegal_op) **else** $error;
- **assert property** (req => done [*->1] ##1 fill [*->1])
  else
  $warning("Fill did not occur for completed mem read. Why?");
- **assert property** (illegal_op)
  else $error("Illegal operation occurred on bus B.");
- **cover property** (req_t1_start)
  begin
  $display("Starting Request t1.");
  end

C.6 Dynamic data within sequences

In addition to matching sequences of events, SystemVerilog includes the ability to sample data within the sequence for later comparison or matching. Both properties and sequences allow us to declare variables that will be assigned data at some point within a sequence.

Note, these variables are separate for each property or sequence start. That is, sequences or properties that can start on each subsequent cycle have independent variables. If shared variables are desired, they should be declared like other RTL variables with code provided to set their values. Here we show the property declaration for variables.

Example C-19 Dynamic data

```
property_declaration ::=
   property property_identifier [ property_formal_list ] ';'
   { property_decl_item }
   property_spec ';'
   endproperty [ ':' property_identifier ]

property_declaration_item ::=
   sequence_declaration  |  variable_declaration

variable_declaration ::=
   list_of_variable_decls_or_assignments
```

Examples include:

```
property addr_sent;
   reg [31:0] addr;
   // Find start and save address, then find the next
   // request and compare the address with the saved one
   (start, addr = new_addr |=> req [*->1]
          ##0 req_addr == addr);
endproperty
```

C.7 Templates

SystemVerilog includes a new macro-like capability (called a template) for encapsulating assertions to allow instantiation in modules and interfaces. These templates can be instantiated like instantiating a module (except that instance names are optional).

Example C-20 Assertion templates

```
template_declaration ::=
  template template_identifier [ '(' template_formal_list ')' ] ';'
    { template_item_declaration }
  endtemplate [ ':' template_identifier ]

template_formal_list ::=
  task_formal_arg { ',' task_formal_arg }

task_formal_arg ::=
  formal_identifier  ['=' expression | event_expr ]

template_item_declaration ::=
  property_declaration
| sequence_declaration
| concurrent_assert_statement
| concurrent_cover_statement

template_instantiation ::=
  template_identifier [instance_name]
   [ '(' list_of_port_connections ')' ] ';'
-- list of port connections include: expression, event control
```

Consider the example template shown below:

Example C-21 Sample template

```
template my_interface(state, clk,
             start,
             stall = 1'b0, // Default value of 0 if not used.
              done);
// A completed sequence starts with start and is followed by
// done 4 cycles later (without any stalls)
  sequence complete;
    (start ##1 !stall [*->4] ##1 done);
  endsequence

  statechk: assert property @(posedge clk)
     (done => state == IDLE);
  stall1: cover property @(posedge clk)
       (start ##1 stall [*->1] within 1 [*5]);
  stall2: cover property @(posedge clk)
       (start ##1 stall [*->2] within 1 [*6]);
  stall3: cover property @(posedge clk)
       (start ##1 stall [*->3] within 1 [*7]);
 endtemplate
```

C.8 System Functions

Assertions are commonly used to evaluate certain specific characteristics of a design implementation, such as whether a particular signal is one-hot. The following system functions are included to facilitate such common assertion functionality:

- $onehot (<expression>) returns true if one and only one bit of expression is high
- $onehot0 (<expression>) returns true if at most one bit of expression is low
- $inset (<test_expression>, <expression> {, <expression>}) returns true if the first expression is equal to at least one of the subsequent expression arguments.
- $insetz (<test_expression>,<expression> {, <expression>}) returns true if the first expression is equal to at least one other expression argument. Comparison is performed using casez semantics, so 'z' or '?' bits are treated as don't-cares and 'x' bits do not match.
- $isunknown (<expression>) returns true if any bit of the expression is 'x'. This is equivalent to:
  "^ <expression> === 1'bx"
- $stable (<expression>) returns true if the previous value of the expression is the same as the current value of the expression.
- $rose (<expression>) returns true if the expression was previously zero and the current value is nonzero. If the expression length is more than one bit, then only bit 0 is used to determine a positive edge.
- $fell (<expression>) returns true if the expression was previously nonzero and the current value is zero. If the expression length is more than one bit, then only bit 0 is used to determine a negative edge.

All of the above system functions have a return type of bit. A return value of 1'b1 indicates true, and a return value of 1'b0 indicates false. These next system functions return a value equivalent to the length of the first (or only) expression.

- $past (<expression>, <num cycles>) returns the value of the expression from *num cycles* ago. If the value did not exist, 'bx is returned.
- $countones (<expression>) returns the number of bits asserted in the expression.

C.9 SystemTasks

SystemVerilog has defined several severity system tasks for use in reporting messages with a common message. These system tasks are defined as follows:

```
$fatal(finish_num [, message
       {, message_argument } ] );
```

This system task reports the error message provided and terminates the simulation with the `finish_num` error code. This system task is best used to report fatal errors related to testbench/ OS system failures (for example, can't open, read, or write to files) The message and argument format is the same as the **$display()** system task.

$error(message {, message_argument });

This system task reports the error message as a run-time error in a tool-specific manner. However, it provides the following information:

- severity of the message
- file and line number of the system task call
- hierarchical name of the scope or assertion or property
- simulation time of the failure

$warning(message {, message_argument });

This system task reports the warning message as a run-time warning in format similar to $error and with similar information.

$info(message {, message_argument });

This system task reports the informational message in a format similar to $error and with similar information.

$asserton(levels, [list_of_modules_or_assertions])

This system task will reenable assertion and coverage statements. This allows sequences and properties to match elements. If a `level` of 0 is given, all statements in the design are affected. If a list of *modules* is given, then that module and modules instantiated to a depth of the `level` parameter are affected. If specifically named assertion statements are listed, then they are affected.

$assertkill(levels, [list_of_modules_or_assertions])

This system task stops the execution of all assertion and cover statements. These statements will not begin matching until reenabled with **$asserton**(). Use the arguments in the same way as $asserton uses them.

$assertoff(levels, [list_of_modules_or_assertions])

This system task prevents matching of assertion and cover statements. Sequences and properties in the progress of matching sequences will continue. Assertion and cover statements will not begin matching again until reenabled with **$assserton**(). Use the arguments in the same way as $asserton uses them.

C.10 BNF

The SystemVerilog BNF represented here is the property specification subset. This subset resides within a module context. It also utilizes the expression BNF subset as part of its expressions.

C.10.1 Use of Assertions BNF:

```
concurrent_assertion_item::=
          property_declaration
        | sequence_declaration
        | template_instantiation
        | template_declaration
        | concurrent_assert_statement
        | concurrent_cover_statement
procedural_assertion_item::=
          immediate_assert_statement
        | concurrent_assert_statement
        | concurrent_cover_statement
```

C.10.2 Assertion statements

immediate_assert_statement ::=

  assert (expression **)** action_block

concurrent_assert_statement ::=

  assert property '(' property_instance **')'** action_block

| **assert property ‘(‘** sequence_instance **‘)’** action_block

| **assert property** ‘(‘ property_spec ‘)’ action_block

concurrent_cover_statement ::=

cover property’(‘ property_instance **‘)’** statement_or_null

| **cover property ‘(‘** sequence_instance **‘)’** statement_or_null

| **cover property** ‘(‘ property_spec **‘)’** statement_or_null

action_block ::=

statement [**else** statement_or_null]

| [statement_or_null] **else** statement_or_null

statement_or_null ::=

statement | ‘;’

C.10.3 Property and sequence declarations

property_declaration ::=

property property_identifier [property_formal_list] **‘;’**

{ property_decl_item }

property_spec **‘;’**

endproperty [**‘:’** property_identifier]

property_formal_list ::=

’(‘ formal_list_item { ‘,’ formal_list_item } ‘)’

formal_list_item ::= formal_identifier [**‘=’** actual_arg_expr]

actual_arg_expr ::= expression | identifier | event_control

property_decl_item ::= sequence_declaration

| list_of_variable_identifiers_or_assignments

sequence_declaration ::=

sequence sequence_identifier [sequence_formal_list] **‘;’**

{ sequence_decl_item }

sequence_spec **‘;’**

endsequence [**‘:’** sequence_identifier]

sequence_formal_list ::=

‘(‘ formal_list_item { ‘,’ formal_list_item } ‘)’

```
sequence_decl_item ::=
        list_of_variable_identifiers_or_assignments.
sequence_spec ::=
        multi_clock_sequence
      | sequence_expr
```

C.10.4 Property construction

```
property_spec ::=
        [ event_control ] property_modifier    property_expr
      | property_expr
property_modifier ::=
        [ disable iff '(' expression ')' ] not
      | disable iff '('expression ')' [ not ]
property_expression ::=
        sequence_spec
      | property_implication

property_implication ::=
        sequence_expr        |-> [ not ] sequence_expr
      | clocked_sequence     |-> [ not ] sequence_expr
      | sequence_expr        |=> [ not ] sequence_expr
      | clocked_sequence     |=> [ not ] sequence_expr
      | multi_clock_sequence |=> [ not ] multi_clock_sequence

multi_clock_sequence ::=
      clocked_sequence { '##' clocked_sequence }
clocked_sequence ::= event_control sequence_expr
event_control ::= '@' event_identifier
      | '@' '(' event_expression ')'
```

C.10.5 Sequence construction

```
sequence_expr ::= [ cycle_delay_range ] sequence_expr
                { cycle_delay_range   sequence_expr }
```

| expression {',' function_blocking_assign } [boolean_repeat]
| **first_match** '(' sequence_expr ')'
| expression **throughout** sequence_expr
| sequence_instance [consecutive_repeat_operator]
| '(' sequence_expr ')' [consecutive_repeat_operator]
| sequence_expr **and** sequence_expr
| sequence_expr **intersect** sequence_expr
| sequence_expr **or** sequence_expr
| sequence_expr **within** sequence_expr
| boolean_expr_op

cycle_delay_range ::=
'##' constant_expression
| '##' **'['** const_range_expression **']'**

const_range_expression ::=
constant_expression **:** constant_expression
| constant_expression **: '$'**

boolean_repeat ::=
consecutive_repeat_operator
| nonconsecutive_repeat_operator
| goto_repeat_operator

consecutive_repeat_operator ::=
'[' '*' const_range_expression **']'**
nonconsecutive_repeat_operator ::=
'[' **'*='** const_range_expression **']'**
goto_repeat_operator ::=
'[' **'*->'** const_range_expression **']'**

boolean_expr_op ::=
sequence_identifier '.' **ended**
| sequence_identifier '.' **matched**
| **$countones** '(' expression ')'
| value_change_functions
value_change_functions::=
$rose '(' expression ')'

| **$past** ‘(‘ expression [‘,’ number_of_ticks] ‘)’
| **$fell** ‘(‘ expression ‘)’
| **$stable** ‘(‘expression ‘)’

C.10.6 Template declaration

template_declaration ::=
template template_identifier
[**‘(‘** template_formal_list **‘)’**] **‘;’**
{ template_item_declaration }
endtemplate [**‘:’** template_identifier]

template_formal_list ::=
task_formal_arg { **‘,’** task_formal_arg }
task_formal_arg ::=
formal_identifier [‘=’ expression | event_expr]

template_item_declaration ::=
property_declaration
| sequence_declaration
| concurrent_assert_statement
| concurrent_cover_statement

template_instantiation ::=
template_identifier [instance_name]
[**‘(‘** list_of_port_connections **‘)’**] ‘;’
-- *list of port connections include: expression, event control*

BIBLIOGRAPHY

[Abarbanel et al. 2000] Y. Abarbanel, I. Beer, L. Gluhovsky, S. Keidar, Y. Wolfsthal, "FoCs--Automatic Generation of Simulation Checkers from Formal Specifications", *Proc. Computer Aided Verification, 12th International Conference, CAV 2000*, pp. 414-427, July 15-19, 2000.

[Accellera OVL 2003] Accellera proposed standard Open Verification Library Users Reference Manual, 2003.

[Accellera PSL-1.0 2003] Accellera proposed standard Property Specification Language (PSL) 1.0, January 2003.

[Accellera SystemVerilog-3.1 2003] Accellera proposed standard SystemVerilog 3.1, April 2003.

[Adir et al., 2002a] A. Adir, R. Emek, E. Marcus. "Adaptive Test Program Generation: Planning for the Unplanned." *Proc. IEEE Int'l High Level Design Validation and Test Workshop*, 2002.

[Adir et al., 2002b] ?A. Adir, R. Emek, E. Marcus. "Adaptive Test Generation." *Proc. Microprocessor Test and Verification Conference*, 2002.

[Alexander 1977] C. Alexander, *A Pattern Language*, New York: Oxford University Press, 1977.

[Alexander 1979] C. Alexander, *The Timeless Way of Building*, New York: Oxford University Press, 1979.

[AMBA-2.0 1999] AMBA^{TM} Specification, Revision 2.0, ARM Limited, 1999.

[Appleton 2000] B. Appleton, "Patterns and Software: Essential Concepts and Terminology", Retrived March 31, 2003, from the World Wide Web: http://www.cmcrossroads.com/bradapp/docs/patterns-intro.html.

[Beizer 1990] B. Beizer, *Software Testing Techniques*, Van Nostrand Rheinhold, New York, second edition, 1990.

[Benjamin et al., 1999] Benjamin, M., D. Geist, A. Hartman, G. Mas, and Y. Wolfsthal. "A Study in Coverage-Directed Test Generation." *Proc. Design Automation Conference*, 1999.

[Bentley 2001] B. Bentley, "Validating the Intel Pentium 4 Microprocessor", *Proc. Design Automation Conference*, pp. 244-248, 2001.

[Bening and Foster 2001] L. Bening, H. Foster, *Principles of Verifiable RTL Design*, Kluwer Academic Publishers, 2001.

[Ben-Ari et al. 1983] M. Ben-Ari, Z. Manna, and A, Pnueli, "The temporal logic of branching time", *Acta Informatica 20*, 1983.

[Bergeron 2003] J. Bergeron, *Writing Testbenches: Functional Verification of HDL Models, Second Edition,* Kluwer Academic Publishers, 2003.

[Betts et al. 2002] J. A. Betts, F. Delguste, S. Brefort, C. Clause, A. Salas, T.Langswert, "The Role of Functional Coverage in Verifying the C166S IP", *Proc. Intn'l Workshop on IP-based System-On-Chip Design,* 2002.

[Borland 2002] "Borland developer network", Article ID: 28432 Publish Date: February 21, 2002 Last Modified: March 06,2002

[Clarke and Emerson 1981] [E. Clarke, E. A. Emerson, "Design and synthesis of synchronization skeletons using branching time temporal logic". *Logic of Programs: Workshop,* LNCS 407, Springer 1981.

[Clarke et al. 2000] [E. Clarke, O. Grumberg, D. Peled, *Model Checking*, The MIT Press, 2000.

[Coonan 2000] J. T. Coonan, "ASIC Design & Verification Top-10 List", Retrived March 31, 2003, from the World Wide Web: http://www.mindspring.com/~tcoonan/asicdv.html.

[Coplien 2000] J. Coplien, *Software Patterns*, SIGS Books and Bultimedia, New York, 2000.

[Devadas et al. 1996] S. Devadas, A. Ghosh, K. Keutzer, "An Observability-Based Code Coverage Metric for Functional Simulation," *Proc. Intn'l Conf. on Computer-Aided Design*, pp. 418-425, 1996.

[Drako and Cohen 1998] D. Drako and P. Cohen, "HDL Verification Coverage", *Integrated System Design*, June, 1998.

[Dwyer et al. 1998] *M. Dwyer, G. Avrunin, J. Corbett*, 2nd Workshop on Formal Methods in Software Practice, March, 1998, Retrived January 30, 2003, from the World Wide Web: http://www.cis.ksu.edu/~dwyer/papers/spatterns.ps.

[e Language Reference Manual] *e Language Reference Manual* is available at https://verificationvalut.com

[Fallah et al. 1998] F. Fallah, S. Devadas, K. Keutzer, "OCCOM: Efficient Computation of Observability-Based Code Coverage Metrics for Functional Verification," *Proc. Design Automation Conference*, pp.152-157, 1998.

[Fitzpatrick et al. 2002] T. Fitzpatrick, H. Foster, E. Marschner, P. Narain, "Introduction to Accellera's assertion efforts", EEDesign, Retrieved June 2, 2002, from the World Wide Web: http://www.eedesign.com/story/OEG20020602S0001

[Floyd 1967] Robert Floyd, "Assigning meanings to programs" *Proceedings, Symposium on Applied Mathematics*, Volume 19, American Mathematical Society, Providence, RI (1967), pp. 19– 32

[FoCs 2003] http://alphaworks.ibm.com/tech/FoCs.

[Foster and Coelho 2001] H. Foster, C. Coelho, "Assertions Targeting A Diverse Set of Verification Tools", *Proc. Intn'l HDL Conference*, March, 2001.

[Foster et al. 2002] H. Foster, P. Flake, and T. Fitzpatrick, "Adding Design Assertion Extensions to SystemVerilog", *Proc. Intn'l HDL Conference*, March, 2002

[Gamma et al. 1995] E. Gamma, R. Helm, R Johnson, J Vlissides, *Design Patterns, Elements of Reusable Object-Oriented Software*, 1995.

[Geist et al.., 1996] Geist, D., M. Farkash, A. Landver, Y. Lichtenstein, S. Ur, and Y. Wolfsthal. "Coveraged Directed Generation Using Symbolic Techniques." *Proc. Int'l Conference on Formal Methods in Computer-Aided Design*, 1996.

[Grinwald et al. 1998] R. Grinwald, E. Harel, M. Orgad, S. Ur, A. Ziv, "User Defined Coverage - A Tool Supported Methodology for Design Verification", *Proc. Design Automation Conference*, 1998.

[Grumberg and Long 1994] O. Grumberg and D. Long. Model checking and modular verification. ACM Transaction on Programming Languages and Systems 16: 843-872, 1994.

[Hoare 1969] C. A. R. Hoare, "An Axiomatic Basis for Computer Programming", *Communications of the ACM* Vol 12, No. 10, October 1969.

[Horgan et al. 1994] J. Horgan, S. London, M. Lyu, "Achieving Software Quality with Testing Coverage Measures," *Computer*, 27(9), pp. 60-69, September 1994.

[IEEE 1076-1993] IEEE Standard 1076-1993 *VHDL Language Reference Manual*, IEEE, Inc., New York, NY, USA, June 6, 1994.

[IEEE 1364-2001] IEEE Standard 1364-2001 *IEEE Standard Hardware Description Language Based on the Verilog Hardware Description Language*, IEEE, Inc., New York, NY, USA, March 2001.

[Kernighan 1978] B Kernighan, *Elements of Programming Style,* McGraw-Hill, 1998

[Keating and Bricaud 2002] M. Keating and P. Bricaud, *Reuse Methodology Manual,* Kluwer Academic Publishers, 2002.

[Kantrowitz and Noack 1996] M. Kantrowitz, L. Noack, "I'm done Simulating; Now What? Verification Coverage Analysis and Correctness Checking of the DECchip 21164 Alpha microprocessor", *Proc. Design Automation Conference*, pp. 325-330, 1996.

[Kripke 1963] S. Kripke, Semantic Considerations on Model Logic. *Proceedings of a Colloquium: Modal and Many valued Logics*, volume 16 of *Acta Philosophica Fennica*, pp. 83-94, August 1963.

[Krolnik 1998] A. Krolnik, Cyrix M3 Phase 1 Report, Cyrix Inc. internal document, 1998

[Krolnik 1999] A. Krolnik, Cyrix M3 Phase 2 Report, Cyrix Inc. internal document, 1999.

[Kroph 1998] T. Kroph, Introduction to Formal Hardware Verification, Springer, 1998.

[Lachish et al. 2002] O. Lachish, E. marcus, S. Ur, A. Ziv, "Hole Analysis for Functional Coverage Data", *Proc. Design Automaction Conference,* 2002.

[Ludden et al. 2002] A. J. M. Ludden, W. Roesner, G. M. Heiling, J. R. Reysa, J. R. Jackson, B.-L. Chu, M. L. Behm, J. R. Baumgartner, R. D. Peterson, J. Abdulhafiz, W. E. Bucy, J. H. Klaus, D. J. Klema, T. N. Le, F. D. Lewis, P. E. Milling, L. A. McConville, B. S. Nelson, V. Paruthi, T. W. Pouarz, A. D. Romonosky, J. Stuecheli, K. D. Thompson, D. W. Victor, and B. Wile, "Functional verification of the POWER4 microprocessor and POWER4 multiprocessor systems", *IBM Journal of Research and Development,* Vol. 46 Num. 1, 2002.

[Meyer 1992] B. Meyer,B. Meyer, Eiffel The Language, Prentice Hall, 1992

[Marschner et al. 2002] E. Marschner, B. Deadman, G. Martin, "IP Reuse Hardening via Embedded Sugar Assertions", *International Workshop on IP SoC Design*, October 30, 2002.

[Moorby et al. 2003] P. Moorby, A. Salz, P. Flake, S. Dudani, T.Fitzpatrick "Achieving Determinism in SystemVerilog 3.1 Scheduling Semantics", *Proceedings of DVCon 2003*, 2003. Retrived March 31, 2003, from the World Wide Web: http://www.eda.org/sv-ec/sv31schedsemantics-dvcon03.pdf

[Nacif et al. 2003] J. Nacif, F. de Paula, H. Foster, C. Coelho Jr., F. Sica, D. da Silva Jr., A. Fernandes, "An Assertion Library for On-Chip White-Box Verification at Run-Time", IEEE Latin American Test Workshop, 2003.

[PCI-2.2 1998] PCI Local Bus Specification, Revision 2.2, PCI Special Interest Group, December 18, 1998.

[Piziali and Wilcox 2002] A. Piziali, T. Wilcox, *Design Intent Diagram: Reasoning About Sources Of Bugs*, Unpublished presentation, 2002.

[Pnueli 1977] A. Pnueli, "The temporal logic of programs". *18th IEEE Symposium on foundation of Computer Science*, IEEE Computer Society Press, 1877.

[Richards 2003] J. Richards, "Creative assertion and constraint methods for formal design verification", *Proceedings of DVCon*, 2003.
FORMAL DESIGN VERIFICATION, Local Bus Specification, Revision 2.2, PCI Special Interest Group, December 18, 1998.

[Riehle and Zullighoven 1996] D. Riehle, H. Zullighoven, *Understanding and Using Patterns in Software Development*, Retrived March 31, 2003, from the World Wide Web:http://citeseer.nj.nec.com/riehle96understanding.html.

[Shimizu et al. 2000] K. Shimizu, D. Dill, A. Hu, "Monitor-Based Formal Specification of PCI," *Proceedings of the Third International Conference on Formal Methods in Computer-Aided Design,* pp. 335-353, November 2000.

[Shimizu et al. 2001] K. Shimizu, D. Dill, C-T Chou, "A Specification Methodology by a Collection of Compact Properties as Applied to the Intel Itanium Processor Bus Protocol," *Proceedings of CHARME 2001*, Springer Verlag, pp. 340-354, 2001.

[Shimizu and Dill 2002] K. Shimizu, D. Dill, "Deriving a Simulation Input Generator and a Coverage Metric From a Formal Specification," *Proceedings of the 39-th Design Automation Conference*, June, 2002.

[Smith 2002] S. Smith, "Synergy between Open Verification Library and Specman Elite", *2002 Proceedings of Club Verification: Verisity Users' Group Meeting*, http://www.verisity.com/resources/vault/index.html

[Sutherland 2002] S. Sutherland, *The Verilog PLI Handbook*, Kluwer Academic Publishers, 2002.

[Tasiran and Keutzer. 2001] S. S. Tasiran and K. Keutzer, "Coverage metrics for functional validation of hardware designs", *Design and Test of Computers*, pp. 36-45, July/August, 2001.

[Taylor et al. 1998] S. Taylor, M. Quinn, D. Brown, N. Dohm, S. Hildebrandt, J. Huggins, J. and C. Ramey, "Functional Verification of a Multiple-issue Out-of-order, Superscalar Alpha Processor—the DEC Alpha 21264 microprocessor", *Proc. Design Automation Conference*, pp. 638-643, June, 1998.

[Ur and Yadin, 1999] Ur, S. and Y. Yadin. "Micro Architecture Coverage Directed Generation of Test Programs." *Proc. Design Automation Conference*, 1999.

[Yuan, et al. 1999] J. Yuan, K. Shultz, C. Pixley, H. Miller, A. Aziz, "Modeling Design Constraints and Biasing in Simulation Using BDDs, " *Proceedings of the IEEE International Conference on Computer Aided Design*, pp. 584-589, November 1999.

[Ziv 2002a] L. A. Ziv, "Using Temporal Checkers for Functional Coverage", *Proc. Intn'l Workshop on Microprocessor Test and Verification,* 2002.

[Ziv 2002b] L. A. Ziv, "Full Cycle Functional Coverage: Coverage Success Stories", *IBM Verification Seminar,* 2002.

[Ziv et al., 2001] Ziv, A, G. Nativ, S. Mittermaier, S. Ur. "Cost Evaluation of Coverage-Directed Test Generation for the IBM Mainframe." *Proc. Int'l Test Conference*, 2001.

Index

Symbols

A

B

C

D

P

Q

R

S

www.vhdlcohen.com